Student Manual

for

Mathematics for Business Decisions

Part 1: Probability and Simulation

Standard Edition: Release 1.5
Alternative Edition: Release 1.5a

Second Edition

Copyright ©2005 by
The Mathematical Association of America (Incorporated)

Library of Congress Catalog Card Number 2005930125

ISBN 0-88385-743-X
Printed in the United States of America

Current printing (last digit):
10 9 8 7 6 5 4 3 2

Student Manual

for

Mathematics for Business Decisions

Part 1: Probability and Simulation

Standard Edition: Release 1.5
Alternative Edition: Release 1.5a

David Williamson, Marilou Mendel, Julie Tarr, and Deborah Yoklic

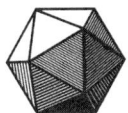

Published and Distributed by
The Mathematical Association of America

CLASSROOM RESOURCE MATERIALS

Classroom Resource Materials is intended to provide supplementary classroom material for students—laboratory exercises, projects, historical information, textbooks with unusual approaches for presenting mathematical ideas, career information, etc.

Committee on Publications

Gerald Alexanderson, *Chair*

Classroom Resource Materials Editorial Board

Zaven A. Karian, *Editor*

William Bauldry	David E. Kullman
Gerald Bryce	Stephen B Maurer
Sheldon P. Gordon	Douglas Meade
William J. Higgins	Edward P. Merkes
Mic Jackson	Judith A. Palagallo
Paul Knopp	Wayne Roberts

101 Careers in Mathematics, 2nd edition edited by Andrew Sterrett

Archimedes: What Did He Do Besides Cry Eureka?, Sherman Stein

Calculus Mysteries and Thrillers, R. Grant Woods

Combinatorics: A Problem Oriented Approach, Daniel A. Marcus

Conjecture and Proof, Miklós Laczkovich

A Course in Mathematical Modeling, Douglas Mooney and Randall Swift

Cryptological Mathematics, Robert Edward Lewand

Elementary Mathematical Models, Dan Kalman

Environmental Mathematics in the Classroom, edited by B. A. Fusaro and P. C. Kenschaft

Essentials of Mathematics, Margie Hale

Exploratory Examples for Real Analysis, Joanne E. Snow and Kirk E. Weller

Fourier Series, Rajendra Bhatia

Geometry From Africa: Mathematical and Educational Explorations, Paulus Gerdes

Identification Numbers and Check Digit Schemes, Joseph Kirtland

Interdisciplinary Lively Application Projects, edited by Chris Arney

Inverse Problems: Activities for Undergraduates, Charles W. Groetsch

Historical Modules for the Teaching and Learning of Mathematics, edited by Victor Katz and Karen Dee Michalowicz

Laboratory Experiences in Group Theory, Ellen Maycock Parker

Learn from the Masters, Frank Swetz, John Fauvel, Otto Bekken, Bengt Johansson, and Victor Katz

Mathematical Connections: A Companion for Teachers and Others, Al Cuoco

Mathematical Evolutions, edited by Abe Shenitzer and John Stillwell

Mathematical Modeling in the Environment, Charles Hadlock

Mathematics for Business Decisions Part 1: Probability and Simulation (electronic textbook), Richard B. Thompson, Christopher G. Lamoureux, and Pamels E. Slaten

Mathematics for Business Decisions Part 2: Calculus and Optimization (electronic textbook), Richard B. Thompson and Christopher G. Lamoureux

Ordinary Differential Equations: A Brief Eclectic Tour, David A. Sànchez

Oval Track and Other Permutation Puzzles, John O. Kiltinen

A Primer of Abstract Mathematics, Robert B. Ash

Proofs Without Words, Roger B. Nelsen

Proofs Without Words II, Roger B. Nelsen

A Radical Approach to Real Analysis, David M. Bressoud

Resources for the Study of Real Analysis, Robert Barbenec

She Does Math!, edited by Marla Parker

Solve This: Math Activities for Students and Clubs, James S. Tanton

Student Manual for Mathematics for Business Decisions Part 1: Probability and Simulation, David Williamson, Marilou Mendel, Julie Tarr, and Deborah Yoklic

Student Manual for Mathematics for Business Decisions Part 2: Calculus and Optimization, David Williamson, Marilou Mendel, Julie Tarr, and Deborah Yoklic

Teaching Statistics Using Baseball, Jim Albert

Understanding our Quantitative World, Janet Andersen and Todd Swanson

Writing Projects for Mathematics Courses: Crushed Clowns, Cars, and Coffee to Go, Annalisa Crannell, Gavin LaRose, Thomas Ratliff, Elyn Rykken

MAA Service Center
P.O. Box 91112
Washington, DC 20090-1112
1-800-331-1MAA   FAX: 1-301-206-9789

Acknowledgements

Initial funding for this manual was provided by Pima Community College's Educational Technology Grant.

The files ***Graphing.xls***, ***Integrating.xls***, and ***Differentiating.xls*** were used in this manual to create many of the graphs and provided solutions to various problems. These files are in the course materials for *Mathematics for Business Decisions Part II: Calculus and Optimization* by Dr. Richard B. Thompson Department of Mathematics University of Arizona and Dr. Christopher G. Lamoureux Department of Finance University of Arizona.

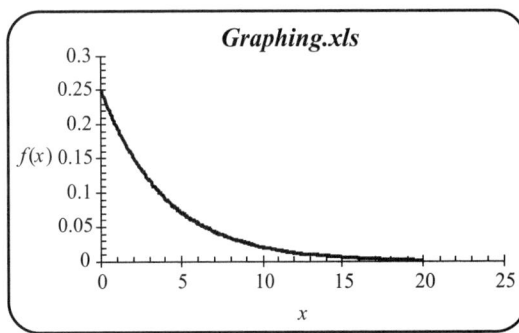

Definition
Formula for $f(x)$
0.25

Computation	
x	$f(x)$
	0.25

Plot Interval	
a	b
0	20

Definition
Formula for $f(x)$
0

Computation	
x	$f(x)$
	0

Plot Interval	
A	B
30	34

Integration Interval	
a	b
32	32.5

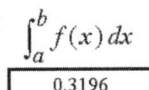

$\int_a^b f(x)\,dx$
0.3196

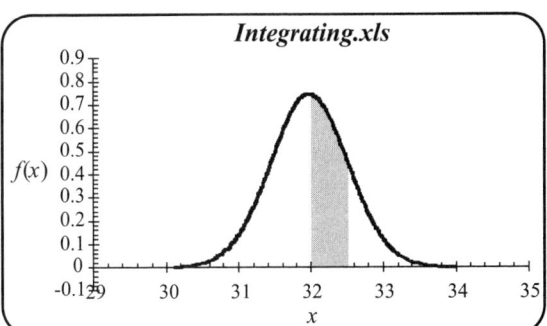

Definition
Formula for $f(x)$
5

Computation		
x	$f(x)$	$f'(x)$
	5	2.000

Plot Interval	
a	b
0	5

Increment
h
0.000001

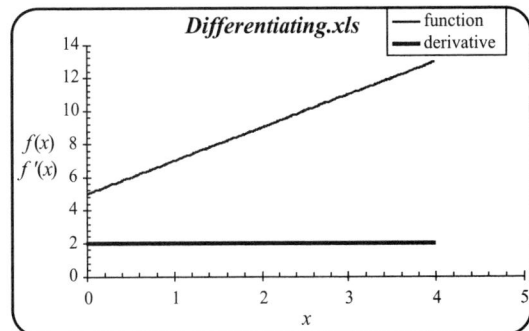

Contents

Acknowledgements vii
Mathematics for Business Decisions: Students and the Workforce 1
Suggestions from Former Business Mathematics Students: What It Takes to be Successful in this Class 3
    Time Management: A Note from the Instructor 5
Having a Successful Team 7
    Bibliography 10
Group Presentations 11
Check List for Group Presentations 17
Technical Writing 19
Technology Tips & Tricks 21
    Basic Microsoft *PowerPoint* Features 21
    Useful Microsoft *Word* Features 24
    Useful Microsoft *Excel* Features 26
Project 1: Loan Work Outs 33
    Business Background 33
    Basic Probability 34
        Properties 34
        Venn Diagrams 36
        Exercises 44
    Summation Notation 51
        Definitions 51
        Exercises 59
    Expected Value 61
        Exercises 65
    Conditional Probability 67
        Definitions 67
        Independent Events 69
        Exercises 73
    Bayes' Theorem 76
        Exercises 81

Project 2/Alternative Project 2 .... **83**

Business Background for Stock Option Pricing 83
Business Background for Managing ATM Queues 87
Compound Interest/Exponential Growth 89
Discrete Compounding 89
Logarithms 92
Continuous Compounding 94
Effective Yield 98
Value of Money 102
Logarithms—Another Look 104
Application 109
Exercises 111
Histograms 118
Exercises 127
Probability Distributions 131
Finite Random Variables 131
Binomial Random Variables 135
Continuous Random Variables 140
Uniform Random Variables 142
Exponential Random Variables 150
Exercises 152
Random Sampling 167
Exercises 177
Simulation 182
Exercises 186

Mathematics for Business Decisions
Students and the Workforce

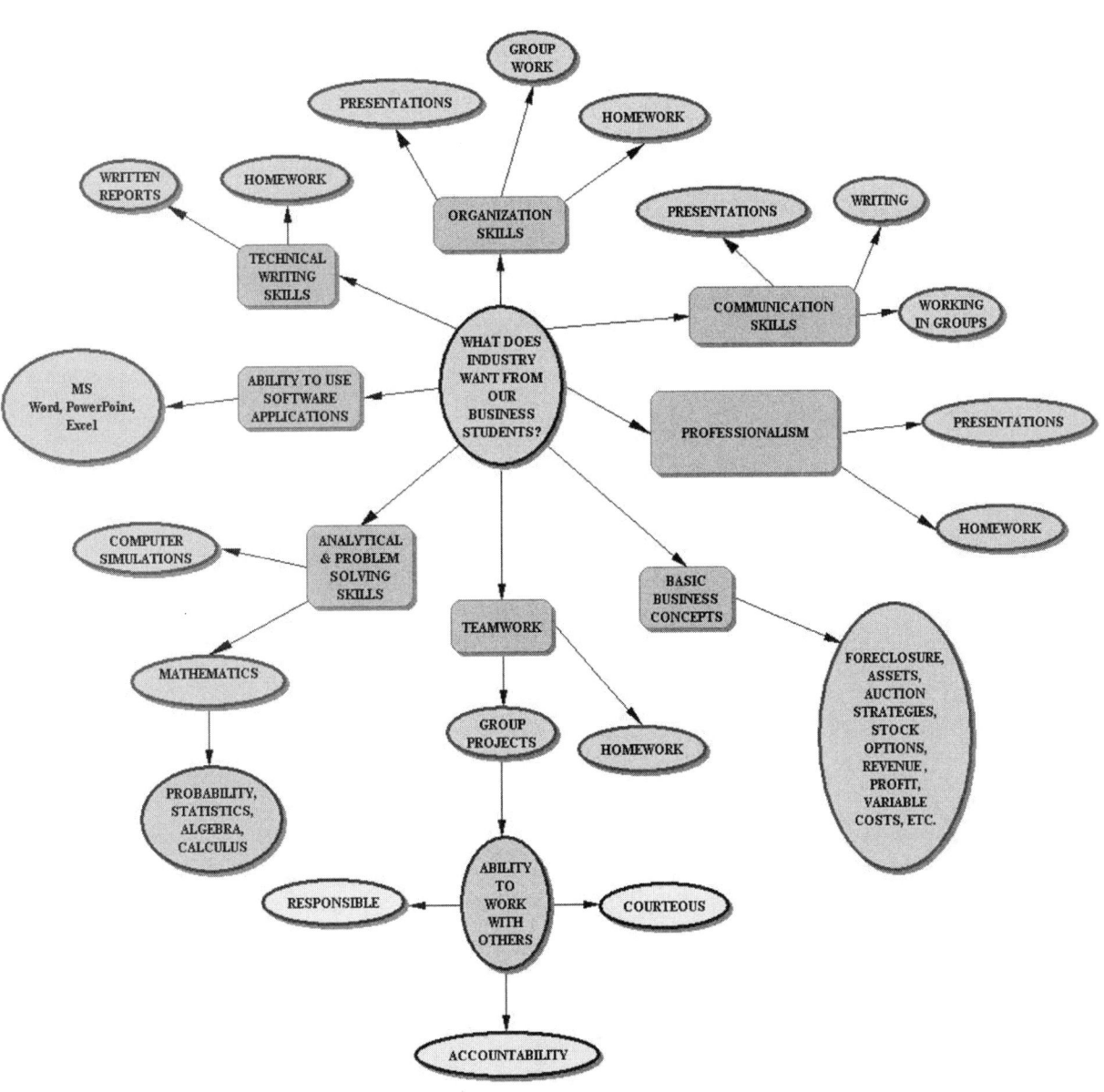

Suggestions from Former Business Mathematics Students: What It Takes to be Successful in this Class

- "Have a very thorough understanding of *Excel*."
- "Work on the project as you go along, so you won't have to rush at the end and you can see how the lecture materials apply to the project. This helps greatly as you write your paper and make your PowerPoint presentation."
- "Stay on top of the reading."
- "Warn students to be prepared to spend a lot of time outside of the classroom working on the team projects."
- "Set aside regular group meeting times each week."
- "Make sure that everyone in the group understands the material before you move on to something new."
- "Working with members of your own group can be helpful in solving problems. Attempting this class on your own is too difficult."
- "Try to get along with your teammates. The projects and the class will be a lot easier if you work well with them."
- "Develop good computer, software, and Internet skills."
- "Attend all lectures and see the teacher at the appointment times. Don't leave questions unanswered."
- "The homework is very complicated, although working with my group made it more understandable and less complicated."
- "The material is not easy – ask for help from other students or your instructor."
- "You need to have a flexible schedule. There will be many times you will need to meet with your group."
- "Look at the examples in PowerPoint. It helped me understand how to do some of the problems using *Excel*."
- "The student should know that going into this class, there would also be times that you meet outside of class. I found it hard to meet with a student that worked full time."
- "The quizzes, tests, and projects have to do with using *Excel*, so you need to have some knowledge on how to use *Excel*."
- "Reading all the assigned texts for the week is crucial."

- "Do not leave everything to the last minute. In this class you can not procrastinate."
- "Collaborate as a group and pull your own weight."
- "Adjust your project to what you have learned in the lectures."
- "This class is very confusing and you need to get help as soon as you are lost."
- "Homework takes a long time. Do not think you can get the homework done an hour before class."
- "Put your project together well before it is due so you can get it reviewed by the instructor."
- "Never allow yourself to feel overwhelmed. The time allowed to complete each project is ample, just do not save it until the end."
- "Understanding the material is the single most important aspect to the course. Taking advantage of the other minds in your group as well as making a point to see your instructor during office hours will benefit you the most."
- "Stay on top of the reading."
- "Group projects should not be put off until the last minute. There are lots of points that need to be covered and they should be done well in advance."
- "Finish the homework at least one class before it is due. That way you will be able to ask your teacher questions that are pertinent to the work."
- "Practice the *Excel* problems. This class is heavily reliant on many different *Excel* functions. Your comprehension with *Excel* may mean the difference between a passing and failing grade on the tests and presentations."
- "While homework can be overwhelming and time consuming, you are cautioned against dividing up the problems among group members. This is because many quiz and test problems are similar to the homework."
- "This class moves very fast and you will probably get behind before you even realize it. Seek help as soon as you need it."
- "Do not overflow your schedule with work and school because it can become dangerous."
- "This is a fast moving class. Do your homework in groups so you can ask questions if you need help."
- "Before meeting with other members of your group to discuss the homework, have everyone do it individually. This way you will be able to compare your answers rather than just relying on those of one person."
- "Make sure that everyone in the group participates and knows how to do each homework problem. Then everyone will be prepared for the test and giving the oral reports."
- "Practice giving your oral reports several times before you actually do them. Also, think of potential questions that your teacher could ask, and make sure everyone in your group knows how to answer them."

- "When working on the projects periodically check your data with the professor and work on it from day one. Waiting until the week before it is due can cause problems."
- "Practice your oral reports so that you can give it without needing to read from anything — you will be less likely to read from an index card or the computer screen. Eye contact is important."

Time Management: A Note from the Instructor

This is not a typical math class. In addition to learning definitions, procedures and working math problems, this course involves group projects, oral presentations, written reports, using *Excel* to solve math problems, and the use of PowerPoint and the equation editor in Word, not to mention an electronic text. All of this requires more of you than a traditional math course. The year you take this course, you will need to make sure you have time in your schedule for these additional demands, particularly that you have time to meet outside of class with your group.

Having a Successful Team

So you've been assigned to Team 2, and the only other person you know in the class has been assigned to Team 1. And, oh, those other people in Team 2! They can't find a time to meet, this one is always on her cell phone, that one has to rush off to work, no one has any idea what the instructor was talking about in class. Oh, you just **know** this is going to be a disaster!

Well, that depends. Depends on **you**, that is. The most important factor in the success of your team is you: your attitude, your expectations, your efforts, your understanding. So here are some ideas to help your group be the best it can be.

Rule #1: Have a Positive Attitude

Go in with a positive attitude, ready to contribute your best to the group. All you really have control over is your own attitude and your own contribution. If everyone has a great attitude, the rest will be that much easier.

What characterizes someone with a "good attitude"? We all know a good attitude when we see one. Someone with a good attitude is curious, eager, interested, well mannered, prepared, helpful, respectful, focused, involved, open-minded, active, a good listener and a team player. Someone with a positive attitude has enthusiasm and optimism for all kinds of ideas. Someone with a positive attitude comes to all group meetings, definitely gets in touch with others when he or she cannot (for a very good reason), has done the readings and the work before the group meets. Have a positive attitude, have a good group.

Rule #2: Be a Scout: Be Prepared

Come to all meetings prepared. Read the electronic text. Work through the examples in the text. If your group works on homework sets together, start all assignments before the group meeting. Go over the Focus Sections and try to figure out how they apply to your group's data. Make a list of questions to bring to the group.

Rule #3: Keep an Open Mind toward Other Group Members

Expect a lot from yourself; expect a little from the others. If you expect the people around you to be something they cannot live up to, you will be disappointed, confused and even angry.

Each person has her/his own work and communication styles, which may be very different from yours. You may have to work with someone you just don't get along with. You have to figure out how to get past whatever clashes come up. Keep in mind that in the "real world" your boss will decide whom you work with. Learn to adapt and make the best of every situation.

Take advantage of each person's strengths. Work together to make up for other's weaknesses. Even that person who seems like a "drag" on the rest of group may turn out to have the best ideas, or the best writing skills, or even be the best mediator. You'll never know if you write them off from the first day.

No one person in the group – not the person with the best grades, nor the person who understands the quickest, nor the person with the best computer skills — not one person has all the information, skills or resources needed for the highest possible quality solution and presentation.

Rule #4: Look at Group Work as an Opportunity

Approach the requirement to work in a group as an opportunity to learn from others. When you teach others, you learn more yourself. Group work is an opportunity to get the best work from each person. One person in the group cannot succeed unless everyone in the group succeeds. You've heard the old saw: the sum is more than its parts. Make that happen!

Rule #5: Nip Conflict in the Bud

Group conflicts often stem from different expectations group members have of each other. At the first meeting, prepare a list of ground rules. These rules might include some of the following: come to meetings on time, come to meetings prepared, turn off cell phones at meetings, outline problems before the group meeting, assign tasks for each group member. Add your own rules to this list! Have *each person* sign the rules.

If — or rather, when — conflict arises, the most important thing to do is **talk the situation over soon**. Talk with the person or persons you have the conflict with, not with others who will inevitably pick sides or with those not even involved. First, though, a couple of minutes of cool-down time might be wise. Whatever you do, don't let things go on too long.

If you don't like someone else's behavior or attitude, talk with them about it in non-accusatory way: "I don't like it when you don't come prepared because then we have to spend time on what the rest of us already know. It would be a better use of our time together solving the problem." Or, "It is hard for me when you come late because I have only an hour and we need all the time together we can manage." Or, "I'm willing try to help you with this, but when I can't help, it would be better for us to go see the instructor than to get angry with me for also not understanding."

Initiate resolution. Be the first person to say you're sorry. Even when you think the other person is wrong, it's not a bad thing to say "I'm sorry you feel that way" or "I'm sorry I offended you in that way."

If you can't agree, then agree to disagree. Compromise: let's do slide #6 your way, and slide #8 my way. The bottom line is: be flexible.

Rule #6: Get Help Before It's Too Late

If there is a concept or homework problem no one in the group understands fully, go to the instructor or a tutor. The instructor might not go over that concept again in class. If you don't know it now, it's sure to show up on an exam, and you still won't know it.

If there is a conflict or problem the group cannot resolve yourselves, go see the instructor before it gets impossible to solve, before the hurt feelings and resentments cannot be turned around. Your instructor may want to meet with you individually and then as a group to try to resolve the conflict.

You are probably in this group for the duration of this project, so it's up to you to make the best of it. It's not helpful to be an ostrich and hope the problems will go away if you ignore them. It is helpful to be that kid with the entrepreneurial stand: when life hands you lemons, make lemonade.

Bibliography

Feenstra, Kristin. "Resolving Conflict in Friendships." *IamNext*. 1 October 2002. <www.iamnext.com/people/tips/conflict.html>.

Felder, R. M. <felder@eos.ncsu.edu> and R. Brent. "Effective Strategies for Cooperative Learning."*Active and Cooperative Learning*. 2001. North Carolina State University. 1 October 2002. <www2.ncsu.edu/unity/lockers/users/f/felder/public/Papers/CLStrategies(JCCCT).pdf>

Godshall, Leslie. "Surviving Group Projects for Life." *IamNext*. 1 October 2002. <www.iamnext.com/academics/inclass/group.html>.

Johnson, Roger T. <johns009@umn.edu> and David W. Johnson. "Cooperative Learning." *Cooperative Learning Center at the University of Minnesota*. 19 June 2002. University of Minnesota. 1 October 2002. <www.clcrc.com/pages/cl.html>.

O'Keefe, Dr. Barbara <b-okeefe@uiuc.edu>. "Teamworks Module: Problem Solving; Lesson 4: Creative Problem-Solving." *Teamworks: Skills for Colaborative Work*. The Sloan Center for Asynchronous Learning Environments (SCALE). 1 October 2002. <www.vta.spcomm.uiuc.edu/PSG/psg14-ov.html>.

Group Presentations

The scenario is that your group is asked to answer questions X and Y for the project.

How would you go about presenting the solution?

Before presenting in front of an audience make sure your group has completed the following tasks:

- **GROUP ORGANIZATION**

    Who will be doing what part of the presentation?

    - **Introduction**
        - State the project objectives.
        - Give any background information.
        - Define the project terms.
        - State the assumptions made.

    - **Steps used to solve questions X & Y**
        - Explain any mathematics used to solve the questions.
        - Explain the *Excel* functions used in the project.
        - Explain why you are doing specific tasks.
        - Use graphs or visual aids.
        - Present solution X.
        - Repeat the steps to answer question Y.

It may take several steps and calculations to reach the final result. Make sure you cover all the steps and calculations.

Remember how you felt lost and confused when you started this project. Your audience will possibly feel that way if your presentation is not clear, and concise, or you skip steps.

If you are not sure about any aspect of your part of the presentation, do not mention it at all because if you do say anything incorrect, you will most likely be asked a question about it.

Now that you have answered all the questions and explained in detail the processes involved for obtaining your solution, your audience has most likely forgotten some, if not all, of your solutions. So it is best to summarize your results and give closure to your presentation.

- **Summary**
  - Restate the solution for X.
  - Restate the solution for Y.
  - Ask the audience if they have any questions.

Before you present in front of your audience make sure the group has practiced together.

- **GROUP PRACTICE**
  - **Practice the entire presentation**
    - Decide who will control the slides.
    - Try to balance the amount of time each person presents.
    - Keep practicing until the presentation goes smoothly.
      - Remember this is a **group** presentation and it only takes one person to make your group look bad.

You should also practice your part of the presentation on your own.

Make sure each group member knows what the other members are talking about. For example, if Steve is talking about how he used *Excel* to solve question X, then all the members should know how to solve question X. Remember you can be asked a question at the end of your presentation about something other than your part of the presentation.

Be prepared for computer problems. The projector may not work or the computer could freeze up. Make sure you have a backup plan. It is a good idea to print your slides so you have a hard copy.

Now that the presentation is organized, your group has practiced and is ready to go, you need to know how to present the material. We need to think about professionalism.

- **PROFESSIONALISM**
  - **Proper attire**
    - No jeans, T-shirts, shorts, sandals, tennis shoes, etc.

Imagine that you are working for a corporation and representing this company. The company wants you to represent them in a professional manner and the first place to start is how you dress.

There is a time and place when it is appropriate to dress up, and when presenting in front of an audience is one of them. Also, the way you dress does not represent your intelligence, but it does reflect some characteristics of your personality and how others may judge you.

For example, laser vision surgery is becoming very popular for near-sighted people, and you are interested in learning more about the procedure, risks involved, recovery time, costs, etc. You have signed up for two seminars.

The first seminar has about 200 potential patients in the audience. The doctor comes out wearing jeans, tennis shoes, and a T-shirt that is not tucked in. The T-shirt is also wrinkled. Think about your first impression.

The second seminar is just as full of potential patients and the doctor is wearing slacks, dress shoes, and a nice shirt. Compare your first impression of this doctor to the first one.

With all other factors considered equal, such as cost, doctor's experience, and the presentation, which doctor would you choose?

An analogy would be going into a restaurant that was not clean. Right away you might leave or at least think that the kitchen is also dirty.

Bottom Line Part of your grade depends on professionalism, and one part of professionalism is how you dress for your presentation. Right now your grade is the only thing being affected, however, later in your career your job or promotion may be affected.

Another part of professionalism is organization. We have already discussed how to organize the group presentation, so what can you do to prepare for your part?

The answer to this question is simple. PRACTICE! PRACTICE! PRACTICE!

■ Organization

- Proofread the PowerPoint slides.
    - Make sure there are no spelling or mathematical errors.
        - Your presentation will not look good if there are errors in your slides. Especially miscalculations, incorrect assumptions, and definitions.
    - Have at least two group members proofread the slides.
        - Would you turn in a writing assignment without proofreading it?
- Practice your part of the presentation for a family member or friend and ask for criticism.
- Practice by yourself.
    - Pretend you are speaking to an audience.
    - Mentally go through your presentation.
- Make cue cards to help guide you through the presentation.
    - Do not read directly from the cue cards.
    - The cue cards should only be used to keep you on track or to help you remember something.
- Understand the material you are presenting.
    - Get help if you are unclear about any aspect of your presentation.
- Present your material systematically.
    - Each part of your presentation should build upon the previous part.
    - There should be a smooth transition when changing speakers.

So far your group has organized and practiced the presentation. You have decided what to wear and practiced your part. At this point you should feel confident about the presentation.

The final step for a professional presentation is the delivery.

- **Presentation delivery**
    - Minimize the number of times you turn your back to the audience.
        - Use your cue cards so you do not have to look at the screen.
    - Face your audience when you speak.
        - Maintain eye contact with the audience.
            - Do not speak to the projector screen.
    - Avoid nervous habits.
        - Do not speak to the audience with your hands in your pockets.
        - Avoid scratching your head, playing with your hair or jewelry, excessive pacing, etc.
    - Do not read directly from the *PowerPoint* slides or cue cards.
        - Since you have practiced, you should be able to explain all the concepts, definitions, and calculations without reading from the cue cards or slides.
        - If there is a long definition, quote, or you completely go blank, then you can read from the cue cards.
    - *PowerPoint* slides should contain bulleted pieces of information.
        - Avoid lengthy definitions or calculations. This accomplishes two things:
            - First, it does not clutter up the slides with a lot of text.
            - Second, it prevents you from reading directly from the slides.
    - Speak clearly
        - Enunciate all words and sentences.
        - Do not use empty filler words such as "uh," "um," or "ok."
        - Avoid speaking in a monotone voice.
            - You may put some of your audience asleep.
        - Do not speak too quickly.
            - You will confuse your audience and eventually loose their attention.
    - Avoid standing in front of the projection screen.
        - While you are presenting to the audience they will be reading from the slides.

    - Format the slides so they are easy to read and follow.
        - Use colors that are clear.
            - Some colors that might look clear on your personal computer may not show up well on the projector.
            - The only way to check this is to view the slides on the projection screen before your group presents. Again, this is a part of being prepared.
            - Do not use dark backgrounds with dark lettering.
    - Do not link to large *Excel* files from *PowerPoint*
        - Use the cut & paste special in *Excel* to show only a part of the spreadsheet.

- All of the spreadsheet is not necessary and it would be difficult for the audience to follow.
  - This prevents the computer from locking up or needing lengthy load times.
    - Points will be deducted from your presentation if this occurs. Remember you have a time limit on the presentation.
- Show graphs and calculations when necessary.

Use the KISS principal— **Keep It Simple Stupid**

Some of the best presentations are the ones that are kept simple and well organized.

HELP The objective of the project is to answer questions X and Y. In order for your audience to understand the process used to answer questions X and Y, the presentation should contain all objectives, definitions, calculations, and graphs. All of these elements should be presented in a logical and clear manner so it is easy for your audience to follow and comprehend the way in which your group has solved the project objectives.

Check List for Group Presentations

Group Organization
- [ ] Identify each group member's part in the presentation.
- [ ] Decide who will control the slides while you are presenting.

Organize the power point slides

Introduction
- [ ] Identify the project objectives.
- [ ] Include any business background information.
- [ ] Define the project terms.
- [ ] State any project assumptions.

Solution process
- [ ] Mathematics used.
- [ ] *Excel* functions used to perform the mathematics.
- [ ] Rationale for your entire solution process.
- [ ] Graphs or visual aids.
- [ ] Present solution.
- [ ] Does your answer make sense?
- [ ] Repeat the process to answer other questions.

Summary
- [ ] Restate all your results.
- [ ] Ask audience if they have any questions.

Professionalism

Organization
- [ ] Practice as a group as well as individually.
- [ ] Use cue cards as a guide.
- [ ] Understand material that is being presented.
- [ ] Present the material in a logical order.
- [ ] Minimize the amount of time you turn your back to the audience.
- [ ] Minimize reading directly from the slides or cue cards.
- [ ] *PowerPoint* slides should contain bulleted pieces of information.
- [ ] Slides should be easy to read.
- [ ] Do not link large *Excel* files into *PowerPoint*.
- [ ] Dress appropriately.

Technical Writing

The technical reports you will be writing in this course will involve detailed explanations as to how your group solved the project. The solutions to the projects involve computer simulation, direct computation, and mathematical concepts and assumptions. The writing style for a technical document is different than the writing style for a non-technical document. To help improve your reports, some guidelines on technical writing can be found at

```
http://www.technical-writing-course.com.
```

This is an excellent resource that is short and easy to following with many examples.

Technology Tips & Tricks

Basic Microsoft *PowerPoint* Features

Getting Started

- Click on ![start]
- Move the pointer to "All Programs" and wait for the sub-menu to appear.
- Move the pointer to the Microsoft Office folder and select Microsoft Office *PowerPoint* 2003.
- Wait for Microsoft *PowerPoint* to open. Your starting window should look similar to the window below. On the right-hand side, click the "X" to the right of "**Getting Started**" to make more room. You can also disable this menu from being automatically displayed by pulling down the "Tools" menu, selecting "Options…," and choosing the "View" tab. Uncheck the "Startup Task Pane."

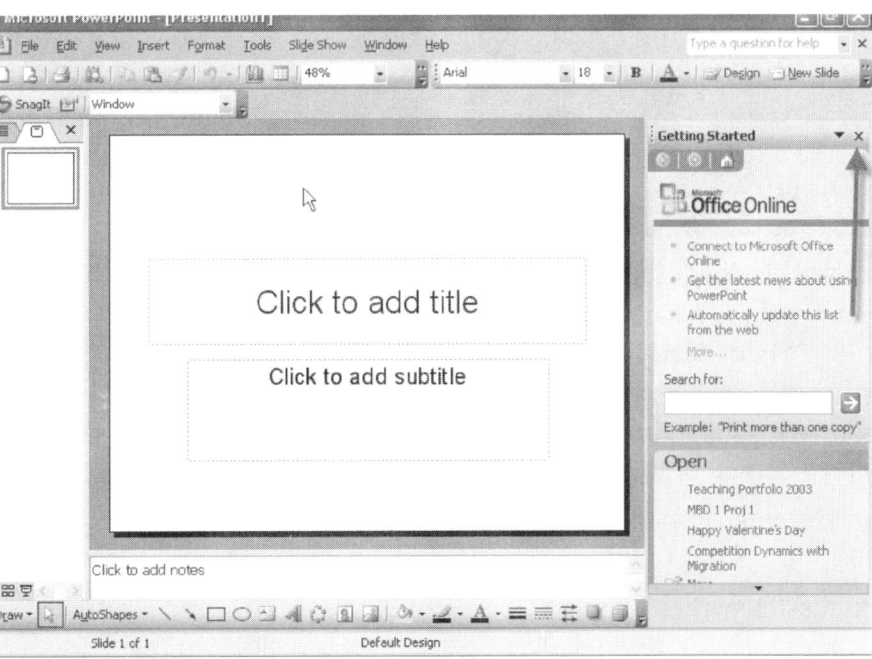

Creating a Basic Presentation

- From the *PowerPoint* toolbar, pull down the "Format" menu and select "Slide Design…"

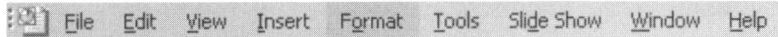

- Choose your design template from the selection that appears on screen. You may also browse for a template if you have purchased additional templates from Microsoft or another vendor.
- Once you have selected a template, pull down the "Format" menu again and this time select "Slide Layout…"
- Select a slide layout for your first slide. You can change your layout for each slide or select "Duplicate Slide" from the Insert pull-down menu for more copies of the same layout.
- Add text (just click and type!) to your slides.
- Don't forget to save your work frequently.

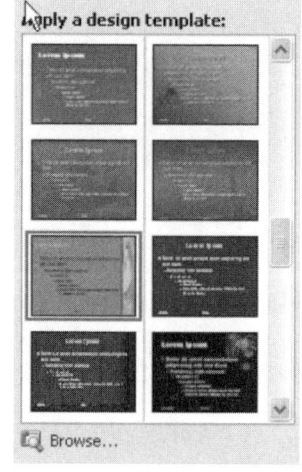

Adding Pictures and Changing the Format or Background

- To add pictures pull down the "Insert" menu and select "Picture."

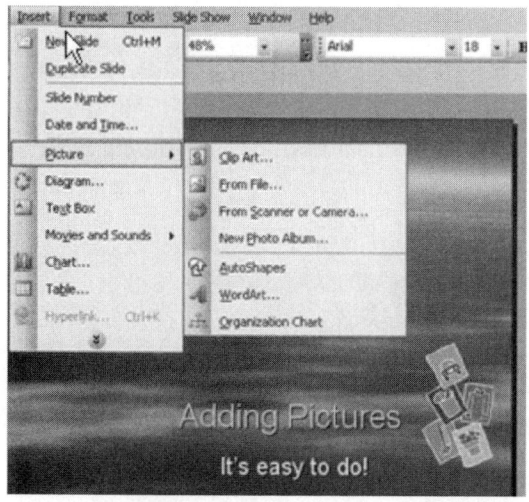

  - Notice that you can add clip art pictures as well as pictures that you have saved as a file (such as a bitmap, gif, or jpeg file).
  - Also note that you can select a layout that has a pre-set spot for a picture or simply insert a picture essentially anywhere you want using the drag-and-drop method.
- The "Format" pull-down menu allows you to change the appearance of the slide.
  - You can change font, background, and line spacing.
  - You can alter slide layouts and slide color schemes.
  - Note that changes can be made to take effect at many levels.
  - If you want to change the font on all of the slides at once, use the "Slide Master" under the "View" pull-down menu.

Adding Links

- Link from one slide to another within a *PowerPoint* presentation, to the Internet, or to your own files.
- Start by selecting the object or text you wish to become a hyperlink.
- Click on the "Hyperlink" option.
- Select another slide in the presentation, input a URL or email address, create a new document, or browse for a file. **Make sure that the file is accessible from where you run your presentation.**
- Warning! Linking large Microsoft *Excel* files can freeze the presentation or take several minutes to load. This is quite embarrassing, especially during timed presentations. If you

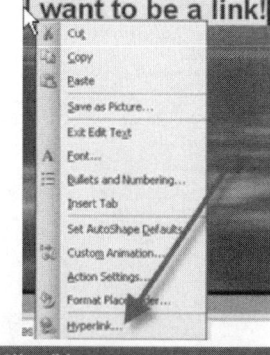

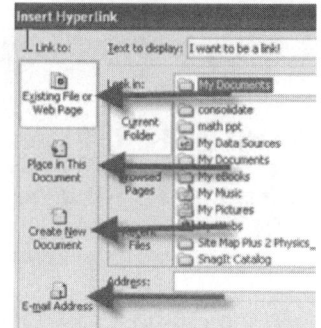

have *Excel* graphs or clips that you wish to show, consider using "Paste Special" under the edit menu and pasting them in as a picture or bitmap instead of as a linked file.

Customizing Animation

- First select which items on the slide you wish to animate.
- From the "Slide Show" pull-down menu, select "Custom Animation."
- Click on the "Add Effect" option in the "Custom Animation" window.
- Once you have chosen your effect, more options for that particular effect will be available, such as speed, direction, and order of animation.
- You can also choose whether you want the animation to occur automatically, after a preset amount of time, or to occur only by mouse-click.
- Warning! Too many animations can make your professional presentation look more like a cheap commercial. Although visuals can spruce up a dry presentation, you should be judicious in your selection of visuals and animations. Remember—not every slide needs something whipping in or wiping out!
- Also, for the most part, it is best to avoid the sound effects. Canned clapping noises may work well in sitcoms, but they usually fall flat in professional presentations.

Printing Options

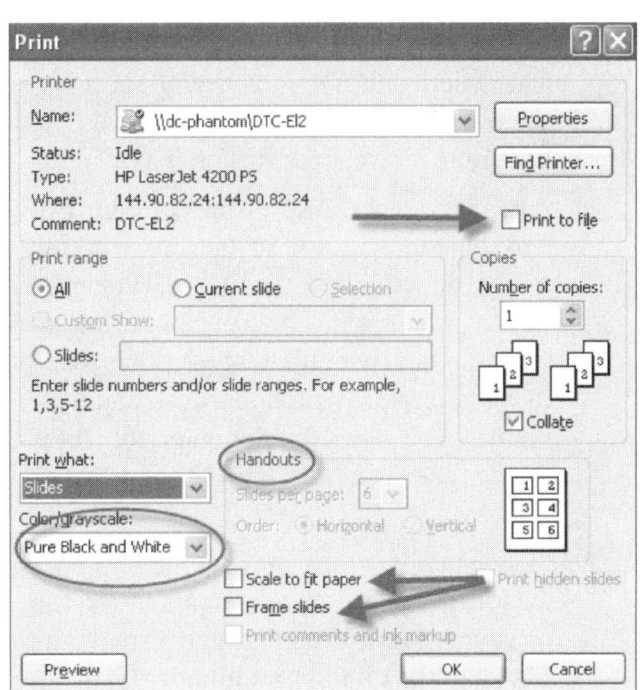

- You can print out all of your slides or just a selection.
- You can print one slide per page or several slides per page or a handout on which your audience can take notes.
- You can print in color, black and white, or by gray-scale. Printing in black and white can save ink and is useful for people with inkjet printers.
- There are numerous other options that you can explore as well, such as framing your slides, scaling them to fit an entire page, and only printing selected slides.

Miscellaneous Pointers

- At the bottom left of your *PowerPoint* work area, you should see a set of icons: . The first icon gives the "normal" view of your work area. This is the view you want when you are building your slides. The second icon is the "slide sorter" view and is useful for organizing your slides, just click, drag, and move a slide wherever you would like it to appear. It is also used for setting your transitions and timings for automated presentations. The third option allows you to preview your presentation.

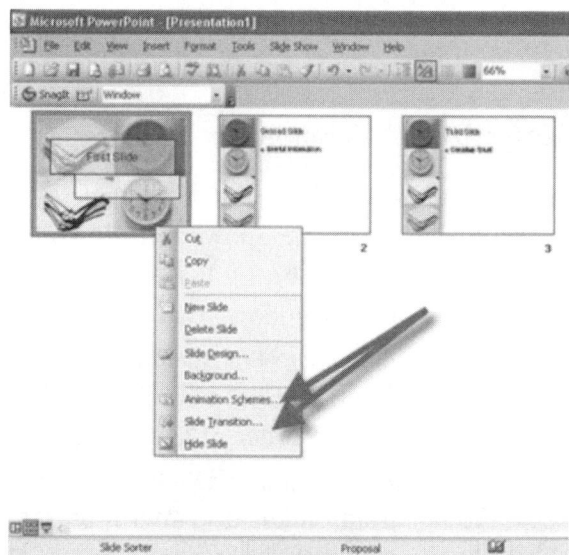

- You can add charts by using the "Add Chart" option under the "Insert" menu, but it is probably better to build your charts using Microsoft *Excel*. Use the "Paste Special" option under the "Edit" menu to insert your charts as bitmaps, as opposed to linked files—remember, you don't want to be standing up in front of an expectant audience with a frozen *PowerPoint* presentation.

- You can add equations in the same way you add them in Microsoft *Word*: Insert-Object-Microsoft Equation, but if you have already typed your equations in *Word*, cutting and pasting will save time.

Useful Microsoft *Word* Features

Using the Microsoft *Equation Editor*

- The Microsoft *Equation Editor* is accessed through the "Insert" pull-down menu by selecting "Object." Once the "Object" window appears, scroll down and select "Microsoft Equation" (whichever version is available).

- After you click "OK" or press Enter, two new objects will appear in your working area. One will be an *Equation Editor* pull-down menu with many categories of mathematics symbols that you may need to word-process your homework and your reports. The other smaller empty object is where the symbols will appear once selected.

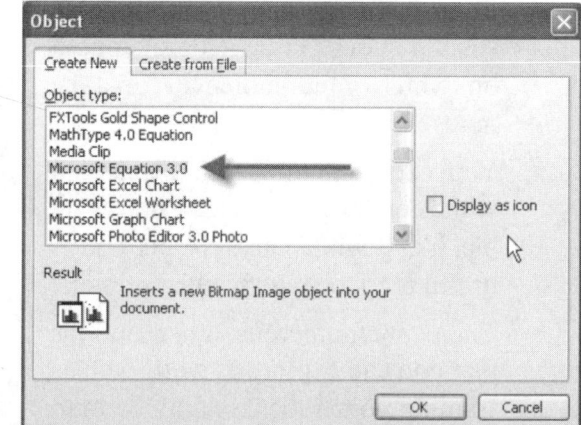

- Notice that when you turn on the *Equation Editor*, the usual *Word* toolbars at the top of your window disappear and a smaller toolbar for use with the *Equation Editor* appears. If you want to align multiple equivalent equations by lining up the equality signs, the "Format" menu is the pull-down menu for you. If you want to insert regular text in a mathematical argument, but do not want to exit the *Equation Editor*, try the

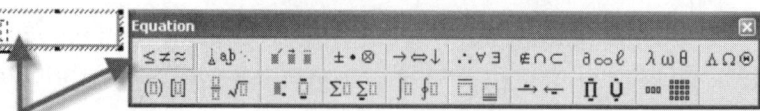

"Style" menu and select "Text"; pull it back down and reselect "Math" to return to the editor's default style. "Other" under the "Size" pull-down menu will allow you to change the size of your math-type font.

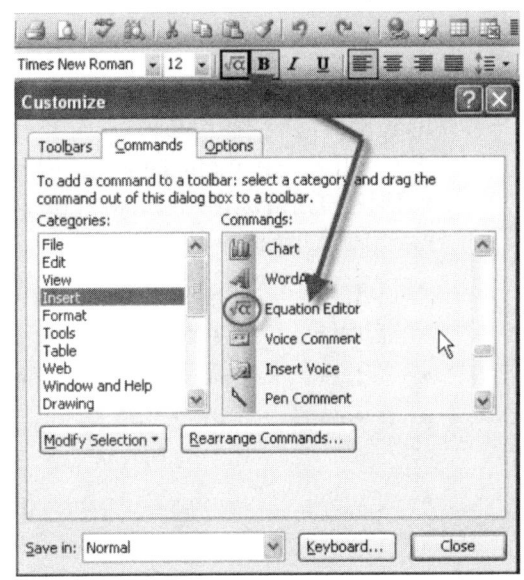

- To save time, you can customize your Microsoft *Word* toolbar to include the Microsoft *Equation Editor* icon. Under the "View" menu on the standard toolbar, select "Toolbars" and scroll all the way down to the bottom to select "Customize". A pop-up window will appear. Select the "Command" tab and "Insert" under the "Categories" option. You will have to scroll down quite a bit under the "Commands" option to find the "Equation Editor" icon (the square root of alpha). Just drag and drop the icon onto your toolbar and you are good to go! Now, you only have to click that icon to run the *Equation Editor*.

Two Special Cases

- Piecewise functions can be a pain! This is especially true when it comes time to word-process them. Suppose you want to word-process the following piecewise defined function:

$$f(x) = \begin{cases} x, & \text{if } x > 0 \\ x^2, & \text{if } x \leq 0 \end{cases}.$$

  A simple way to proceed is as follows:

  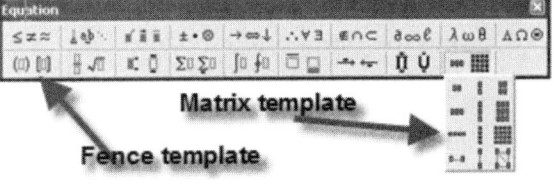

  - Call up the equation editor,
  - Type in "$f(x) = $ " using the keyboard,
  - Select the single left bracket from the "Fence templates" (the pull-down menu on the far bottom left),
  - Select the 4-by-4 matrix from the "Matrix templates" on the far right,
  - Insert the four entries, x, x^2, $x > 0$, and $x \leq 0$, each in their own cell.
  - Voilà! You have a perfectly aligned piecewise function.

- Sample spaces can also provide word-processing challenges. Let's walk through word-processing the sample space:

$$\begin{Bmatrix} (1,1) & (1,2) & (1,3) \\ (2,1) & (2,2) & (2,3) \\ (3,1) & (3,2) & (3,3) \end{Bmatrix}$$

  - Call up the equation editor,
  - Select the double brackets, "{ }" from the "Fence templates," and

- Select the 3-by-3 matrix from the "Matrix templates" on the far right (you can construct higher than a 4-by-4 matrix by selecting the bottom-corner rectangular matrix of unspecified size).

Hotkeys

- Sometimes pull-down menus can be tedious to use. You must interrupt your keyboard usage, find your mouse, click here, click there, ..., sigh. Here are some shortcuts to avoid the pull-down menus in the *Equation Editor* for some commonly used objects (Note: **CTRL+Character** indicates that you should simultaneously depress both the control key and that character):
    - **Ctrl+F**   inserts the fraction builder.
    - **Ctrl+/**   inserts the in-line slash fraction builder.
    - **Ctrl+9**   inserts parentheses.
    - **Ctrl+H**   allows you to type in superscripts (**Tab** puts you back on the main line, as do the arrow keys).
    - **Ctrl+L**   allows for subscripts (**Tab** puts you back).
    - **Ctrl+R**   inserts a square root builder.
    - **Ctrl+I**   inserts an definite integral builder.
- Here are a few more useful shortcuts in which you first simultaneously depress the Control key and "T", let go, and then depress another character. This is represented as **Ctrl+T, character**.
    - **CTRL+T, S**   inserts the summation notation builder
    - **CTRL+T, M**   inserts a 3×3 matrix
    - **CTRL+T, U**   inserts an underscript, which is very useful with limit notation.
- Look for more shortcuts by visiting the *Equation Editor* Help menu and typing "shortcuts" into the index dialogue box.

Templates

- When creating reports, working from a pre-structured template can save time. Under the "File" menu, select "New" to bring up an option window on the right-hand-side of your workspace. Select either "Templates on Office Online" or "On my computer..." for free report templates to help you get started on the path to professional looking reports.
- Make sure to modify the report somewhat to give it your team's personal touch.

Useful Microsoft *Excel* Features

General Helpful Hints

- Highlighting a cell reference in a formula and depressing the **F4** function key will automatically render the reference absolute; e.g., **F4** will change a cell reference such as A1 to A1.
- If you have set the calculation capabilities to manual (through the "Tools" pull-down menu and "Options..." calculation tab), press the **F9** function key to instigate calculations.

You should also uncheck "recalculate before saving" so *Excel* does not change the values.

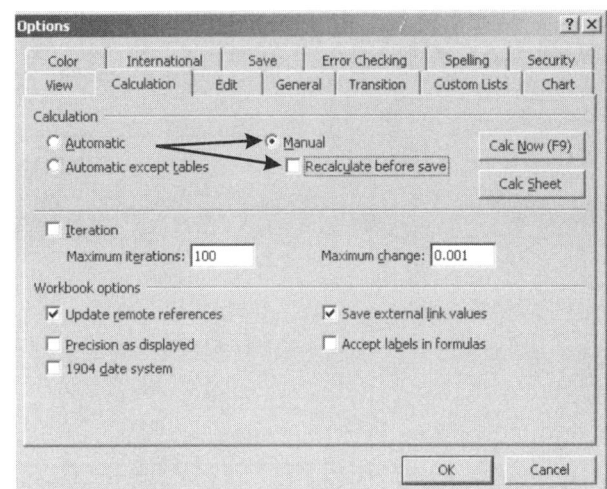

- Using the **CTRL+SHIFT** and the arrow keys (as well as the END and HOME keys) help select large ranges of data instead of scrolling with the mouse, since it is often hard to control the scrolling speed.

- The **Go To** command (under the "Edit" pull down menu) is also useful for navigating within large spreadsheets.

- Under the "Edit" pull-down menu, there is a "Fill" sub-menu. Choose "Series" and a dialogue box will appear. Quickly generating a column of values with a prescribed start and stop point will be useful for many problems in *Mathematics for Business Decisions*! The "Series" dialogue box shown will create the column of numbers starting at one, ending at five thousand, in steps of size one.

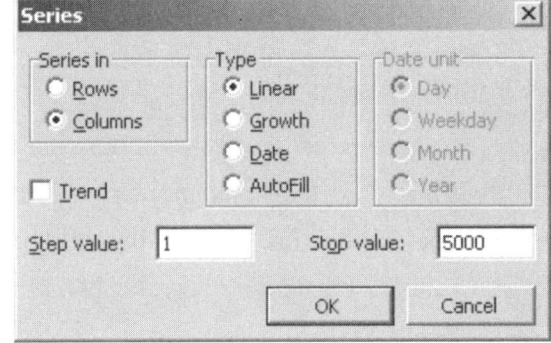

- *Right-clicking* on an object usually gives options designed particularly for that object.

- Right-click on a chart and select *Chart Options* to change the chart's title, gridlines, axes, data labels, and move the legend. Select *Format Chart Area* to format the background area and font used. *Source Data* allows you to change the range of data selected, add more graphs, or detail the legend. *Chart Window* is nice if you want a full-page printout of your graph.

- Right-click in the plot area itself to format the plot area; on the curve to format the curve or add a trend line to a scatter plot; on the axis to change the scale, etc.

- Order of operation is just as important when using Microsoft *Excel* as it is with any calculation tool. You may know that $\frac{x+1}{x+8}$ means divide the sum $x + 1$ by the sum $x + 8$ but, in *Excel*, you input formulas in-line. Thus $x+1/x+8$ will be interpreted by *Excel* as $x+(1/x)+8$ or $x+\frac{1}{x}+8$. Use parentheses to avoid such misinterpretations. $\frac{x+1}{x+8}$ should be input as $(x + 1)/(x + 8)$.

- Similarly, 5^{x+1} typed in-line becomes $5^\wedge(x + 1)$ *not* $5^\wedge x + 1$ The latter expression is equivalent to $5^x + 1$.

Making Graphs with *Excel*

From scratch: Remember the days when you graphed by plotting points? Well, they're back! The function below was created in Microsoft *Excel* with the **Scatter Plot** chart type in the

Chart Wizard. The graph of $f(x) = \begin{cases} 0 & \text{if } x < 0 \\ e^{-x/0.5} & \text{if } x \geq 0 \end{cases}$ is given below.

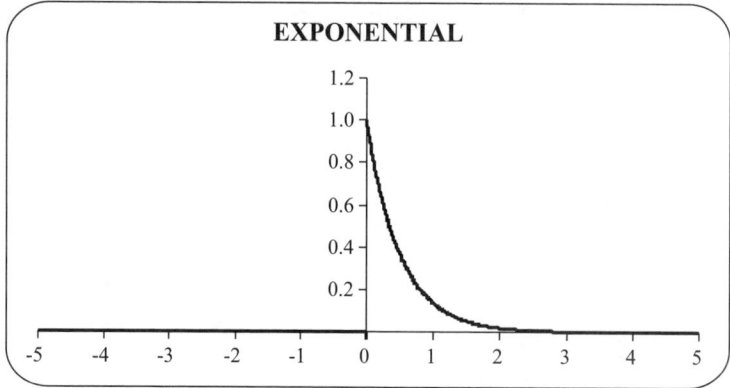

Here are instructions to build the above graph using Microsoft *Excel*:

- You must first choose the interval, [*a*, *b*], over which you wish to graph your function and your step size, Δx. Do not make Δx too big or your scatter plot won't look continuous. In the graph above, $a = -5$, $b = 5$, and $\Delta x = 0.01$.

- In your *Excel* worksheet, construct headings for the *x*- and *y*-columns.

- Input your left-hand endpoint, *a*, in your *x*-column.

- Pull up the "Fill Series" dialogue box from the "Edit" menu.

- Input Δx as your "Step value" and input your right-hand endpoint, *b*, as your "Stop value".

- Select "OK."

- Select the cell adjacent to your first *x* value and type "=" to activate the formula editor.

- Type in the formula for the function you wish to graph. You will need to use relative cell referencing instead of *x* for your input values. This is automatically accomplished by clicking in that cell. Be sure to enter the formula before clicking in any other cell as this action will disturb your formula, and yourself.

- For the function graphed, the necessary *Excel* function is "=IF(A2<0,0,EXP(-A2/0.5))." This formula assigns a value of zero to *y* if the *x* value is also less than zero, otherwise $y = e^{-x/0.5}$. Note that EXP(*x*) in *Excel* is equivalent to e^x.

- The logical "IF" function is a great help in creating piecewise functions. Read through Microsoft *Excel* Help for more details on using the logical "IF" function.

- Generate the entire column of output values at once by double-clicking the fill handle in the bottom right corner of the cell in which you input the formula.

- Use the *Chart Wizard* and choose the Scatter Plot chart type and select the picture of disconnected dots under the chart sub-type. Click the "Next" button at the bottom of the *Chart Wizard* dialogue box and select the "Series" tab. Cell reference your *x*-values in the box labeled "X Values" and cell reference your *y*-values in the box labeled "Y Values." Either select "Finish" to exit the wizard or "Continue" to put some finishing touches on your graph, such as gridlines, labels, and a legend.

A somewhat more complicated example follows:
- Uniform Distributions have graphs that come in three pieces. So the "If, then, else" logical structure illustrated above must be modified.
- Consider the *p.d.f.*, the probability density function, of a uniform random variable, X, on the interval $[0, 10]$. The *p.d.f.* of X is

$$f(x) = \begin{cases} 0 & \text{if } x < 0 \\ \dfrac{1}{10} & \text{if } 0 \le x \le 10. \\ 0 & \text{if } x > 10 \end{cases}$$

- The graph of *f* can be generated using an embedded "If, then, else if" structure as follows:
    - Select a starting value, such as -5, then use *fill series*, as described in the previous example, to generate a column of closely spaced input values. Choose a *stop value* that covers the major features of the graph, such as 15.
    - When you type in the formula this time, however, you must use an *If Statement* within the *If Statement* that you are building; e.g., an *embedded if*. See the diagram to the right.

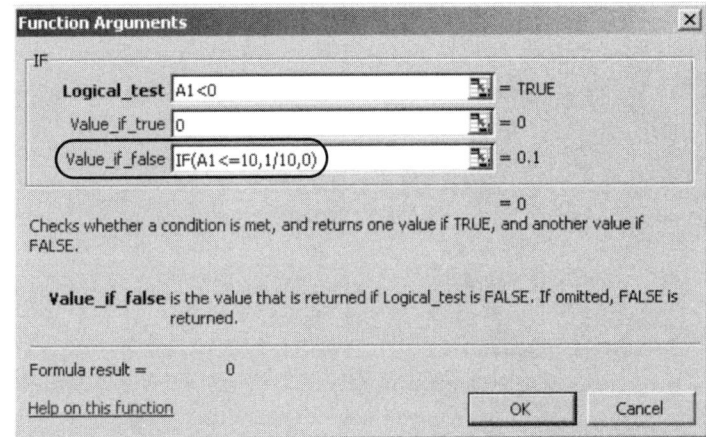

    - As before, use the *fill handle* to copy this formula down the entire column, and the *Chart Wizard Scatter Plot* to create the graph. See the graph of this function below.

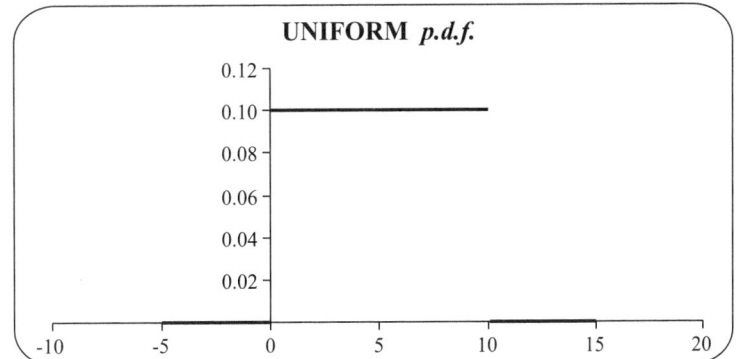

Conditional Formatting

Conditional formatting can not only help spruce up your Microsoft *Excel* files for reporting purposes, and it can also help you notice important calculations. For example, suppose that you are using *Excel* to monitor complicated budgets for a department that you are managing. As your units send in their numbers, your assistant updates your files and sends them over to you for quick check. Wouldn't it be nice if overspent accounts were automatically formatted in red and accounts dangerously close to going over their allocated funds were automatically formatted in some other color, perhaps a dangerous looking shade of pink? Suppose Division #1's budget is not to exceed $5,000.00 per quarter and Division #5 is not to exceed $10,000.00 per quarter. Also, suppose that all divisions together should not exceed $25,000.00 per quarter and $100,000.00 per year. You have just received the following budget update:

Department Budget					
	1st Quarter	2nd Quarter	3rd Quarter	4th Quarter	Division Totals
Division#1	$4,998.00	$4,789.00	$5,023.00	$4,920.00	$19,730.00
Division#2	$6,200.00	$6,298.00	$6,100.00	$5,987.00	$24,585.00
Division#3	$1,852.00	$1,997.00	$1,856.00	$1,945.00	$7,650.00
Division#4	$1,978.00	$2,100.00	$2,200.00	$2,178.00	$8,456.00
Division#5	$9,900.00	$9,999.00	$9,789.00	$9,745.00	$39,433.00
Quarter Totals	$24,928.00	$25,183.00	$24,968.00	$24,775.00	$99,854.00

Granted, this is a small department and an extremely simplified budgetary view, but it would still take a moment or two to find the trouble-makers. This manual is not in color, so both red and hot pink are out, but we can still make the problems quickly stand out automatically through conditional formatting. Shoppers, compare the above boring and unformatted chart, to the following chart:

Department Budget					
	1st Quarter	2nd Quarter	3rd Quarter	4th Quarter	Division Totals
Division#1	$4,998.00	$4,789.00	*$5,023.00*	$4,920.00	$19,730.00
Division#2	$6,200.00	$6,298.00	$6,100.00	$5,987.00	$24,585.00
Division#3	$1,852.00	$1,997.00	$1,856.00	$1,945.00	$7,650.00
Division#4	$1,978.00	$2,100.00	$2,200.00	$2,178.00	$8,456.00
Division#5	$9,900.00	$9,999.00	$9,789.00	$9,745.00	$39,433.00
Quarter Totals	$24,928.00	*$25,183.00*	$24,968.00	$24,775.00	$99,854.00

Okay, it is not that much more exciting, but the dollar amounts that have exceeded their respective allocations are more easily detected. Conditional formatting allows the user to specify formatting properties dependent upon conditions specified on the contents of a cell. In the above example, the cells across Division #1 were formatted to use a bold italic font style and grey cell fill if the dollar amount exceeded the budgetary quarterly limit for Division #1 of $5,000.00. Similarly, the quarter total cells were formatted in the same way if a quarterly total exceeded the limit of $25,000.00.

Closer inspection shows that, on at least a couple of occasions, certain divisions came dangerously close to their limits. We can add additional criteria to highlight these dangerously close situations.

Department Budget					
	1st Quarter	2nd Quarter	3rd Quarter	4th Quarter	Division Totals
Division#1	$4,998.00	$4,789.00	*$5,023.00*	$4,920.00	$19,730.00
Division#2	$6,200.00	$6,298.00	$6,100.00	$5,987.00	$24,585.00
Division#3	$1,852.00	$1,997.00	$1,856.00	$1,945.00	$7,650.00
Division#4	$1,978.00	$2,100.00	$2,200.00	$2,178.00	$8,456.00
Division#5	$9,900.00	$9,999.00	$9,789.00	$9,745.00	$39,433.00
Quarter Totals	$24,928.00	*$25,183.00*	$24,968.00	$24,775.00	$99,854.00

Now cells that contain dollar amounts within $50.00 of their respective limits stand out as well. For your spreadsheets, you should probably use colors as opposed to patterns. Cell formatting is easily accomplished by performing the following steps:

- Under the "Format" pull-down menu, select "Conditional formatting…"; you may have to use the menu expansion arrows at the bottom of the menu.
- The dialogue box below will appear. Input the desired conditions with regard to the contents of the cell, click the "Format…" button, chose your conditional formatting options, and select "OK."
- As in the above example, you may have more than one condition for which you want to impose a conditional formatting protocol. Notice the "Add>>" button at the bottom of the dialogue box? Yes, it really is as easy as that! Add multiple conditions using the same or different associated formatting options.

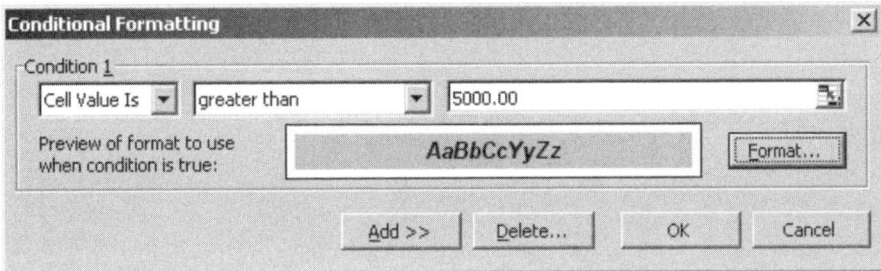

- Note that the full set of formatting options is not available through conditional formatting, but you should have enough options to make your very good, very bad, and somewhat worrisome values all stand out.

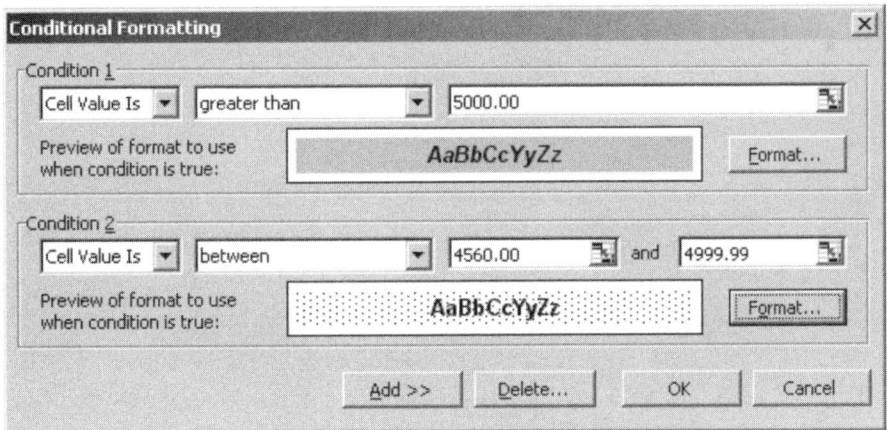

Project 1: Loan Work Outs

Business Background

The primary role of commercial banks in the United States is to facilitate the transfer of money between borrowers and lenders. Individuals, businesses, and governments with excess funds deposit those monies into the local bank in return for interest on the deposits. In turn, the bank profits by lending those monies to individuals, businesses, and governments who are in need of funds.

Loans made by commercial banks are generally of two types — consumer loans and commercial loans. Consumer loans are loans that are made to individuals, and commercial loans are loans that are made to businesses or governments. Loans of either type may be revolving or term loans. The proceeds of revolving loans become available again immediately upon repayment; the proceeds of term loans are available for one-time use only.

Individuals may borrow funds for short-term expenses or for the purchase of assets such as appliances, cars, or homes. The lending decision for a consumer loan is generally based upon the residential, employment, and credit histories of the borrower. In addition, an appraisal is required when an asset is used as collateral. The bank retains title to the asset until the loan is repaid so that the bank can recover the funds from the sale of the asset if the borrower fails to repay the loan. A revolving loan typically requires monthly payments where the amount of each payment is equal to the larger of:

1. A fixed percentage of the average daily balance, and
2. A fixed amount.

A term loan is generally repaid in monthly installments consisting of both principal (reduction of the outstanding balance) and interest.

Corporations may borrow funds for short-term expenses, to finance short-term assets such as inventory or accounts receivable, or for the purchase of long-term assets such as property or equipment. The lending decision for a commercial loan is generally based upon an analysis of the company, an analysis of the industry in which the company operates, and an analysis of the economy. These analyses generally include:

1. Examination of the company's financial statements.
2. Computation of projected cash flows.
3. Appraisal of the company's assets.
4. Evaluation of the proficiency of the company's management.
5. Comparisons of the company's performance and financial condition to that of other companies in the same industry.
6. Examination of the industry life cycle.
7. Assessment of the condition of the general economy.

Commercial loans require monthly interest payments and repayment of the amount borrowed on the maturity date of the loan.

If a borrower fails to make a scheduled loan payment, then the borrower is said to be in default. The bank must then decide how to proceed. This decision requires further analysis and is not made lightly. One alternative is to immediately foreclose on the loan. This often requires liquidation of assets and may put the company out of business. If the bank chooses this option, then it will receive the proceeds from the sale of the assets. This amount is called the foreclosure value and is typically much less than the amount owed. Another alternative is to work out the loan. This requires renegotiation of the loan agreement. If the borrower remains in default, then the bank will eventually receive the proceeds from the liquidation of the business. This amount is called the default value and is typically much less than the foreclosure value due to further depreciation in the value of the company's assets. On the other hand, if the work out is successful, then the bank will receive the full value of the loan. This information is summarized in the following diagram.

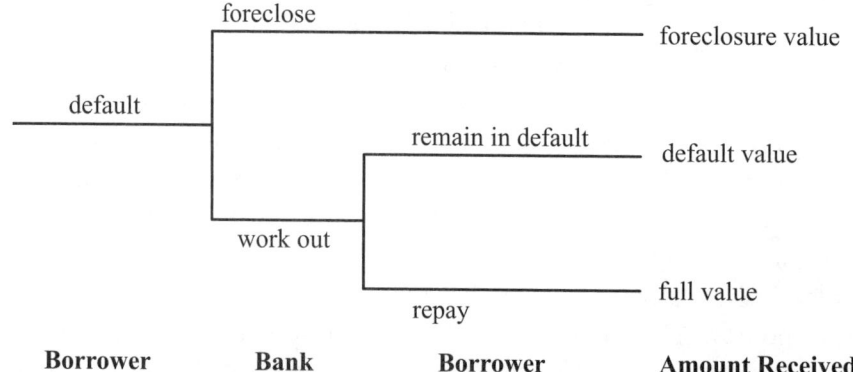

Please visit http://www.aba.com for additional information about the commercial banking industry.

Basic Probability

Properties

The probability that something happens is the likelihood that it occurs. We can almost certainly agree that the true probability of getting a head when we toss a fair coin is ½. If we toss a coin 10 times, we might not get five heads out of the ten tosses. But if we toss the coin 100 times, the number of heads will be approximately 50. If the number of tosses is n, the larger n is, the closer the number of heads is to $n/2$. So one way of determining the probability of something happening is the following:

$$\text{Probability something happens} = \frac{\text{number of times that something happened}}{\text{total number of repetitions of the experiment}}.$$

The more times you repeat the experiment, the closer this experimental, or empirical, probability will be to the theoretical probability. Often in practical situations, the best we can do is estimate the theoretical probability with the empirical probability.

Consider the experiment of tossing a six-sided die. We record whether the number of spots on the top face is "1" or some other number. Assuming the die is fair, we expect any one of the six

Basic Probability

faces to show up about the same number of times as any other face. When this happens, we say that the outcomes are *equally likely*. There are six possible outcomes, only one of which is a "1" and five of which are not a "1." So the probability that the number of spots on the top face is "1" is 1/6, and the probability that the number is some other number is 5/6. Notice that the two probabilities add to 1.

This illustrates the following important facts about probabilities:

1. Every probability must be a number between 0 and 1.
2. The sum of all the probabilities associated with an experiment must add up to 1.
3. If all the n outcomes of an experiment are equally likely, then the probability of any one outcome is $1/n$.

Example 1: Which of the following probabilities are feasible for an experiment having a sample space with three outcomes $\{s_1, s_2, s_3\}$?

    a. $P(s_1) = 0.4$, $P(s_2) = 0.4$, $P(s_3) = 0.4$
    b. $P(s_1) = 0.5$, $P(s_2) = 0.7$, $P(s_3) = -0.2$
    c. $P(s_1) = 0.25$, $P(s_2) = 0.5$, $P(s_3) = 0.25$

SOLUTION

    a. This is not feasible, since the sum of the probabilities is larger than 1.
    b. This is not feasible, since $P(s_3)$ is negative; all probabilities must be between 0 and 1, inclusive.
    c. This is feasible; each probability is between 0 and 1, and the probabilities add to 1.

Example 2: Suppose there are two urns, Urn I and Urn II, each containing white balls and red balls. An experiment consists of selecting an urn, selecting a ball from that urn, and noting its color.

    a. What is a suitable sample space for this experiment?
    b. Describe the event "Urn I is selected" as a subset of the sample space.

SOLUTION

    a. Let *I* be the event Urn I is selected, *II* be the event Urn II is selected, *W* be the event a white ball is drawn, and *R* be the event a red ball is selected. There are four outcomes: Urn I, White Ball; Urn II, White Ball; Urn I, Red Ball; Urn II, Red Ball. We can describe these as follows: $\{IW, IIW, IR, IIR\}$
    b. If "Urn I is selected," the ball drawn can be either white or red. So there are two outcomes in the event "Urn I is selected," namely $\{IW, IR\}$.

Example 3: Suppose a red die and a green die are tossed and the numbers of spots on the top faces are observed.

    a. What is the probability that the numbers add up to 8?
    b. What is the probability that the sum of the numbers is less than 5?

SOLUTION

    a. There are five outcomes for which the numbers add up to 8:
$$\{(2,6), (3,5), (4,4), (5,3), (6,2)\}.$$

There are 36 possible outcomes when two dice are tossed. Since each of the 36 outcomes is equally likely, the probability that the numbers add to 8 is 5/36.

$$S = \begin{Bmatrix} (1,1) & (1,2) & (1,3) & (1,4) & (1,5) & (1,6) \\ (2,1) & (2,2) & (2,3) & (2,4) & (2,5) & (2,6) \\ (3,1) & (3,2) & (3,3) & (3,4) & (3,5) & (3,6) \\ (4,1) & (4,2) & (4,3) & (4,4) & (4,5) & (4,6) \\ (5,1) & (5,2) & (5,3) & (5,4) & (5,5) & (5,6) \\ (6,1) & (6,2) & (6,3) & (6,4) & (6,5) & (6,6) \end{Bmatrix}$$

b. There are six outcomes for which the numbers add up to something less than 5, namely to 2, 3, or 4: {(1,1), (1,2), (1,3), (2,1), (2,2), (3,1)}.

So the probability that the numbers add to something less than 5 is 6/36 = 1/6.

Let's define the event E to be "the vehicle is red in color." Then the event "not E" is "the vehicle is not red." The event "not E" is known as **E complement**; the notation used is E^C, where the C is a superscript.

If E is the event a 2 shows up when you toss a 6-sided die, then E^C is the event 1, 3, 4, 5, or 6 shows up.

If E is the event a voter cast his/her ballot for George W. Bush, then E^C is the event the voter cast his/her ballot for some other candidate.

We can think of E and E^C as being the only two possible outcomes of an experiment. Since we know the sum of all the probabilities must be equal to 1, we have $P(E) + P(E^C) = 1$ or $P(E^C) = 1 - P(E)$.

Example 4: If the event E is that a worker is female, what is the event E^C?

SOLUTION

E^C is the event a worker is not female.

Example 5: The weather forecaster predicts that there is a 20% chance of rain tomorrow. What is the probability it does not rain?

SOLUTION

By complements, $P(\text{no rain}) = 1 - P(\text{rain}) = 1 - 0.20 = 0.80$.

Venn Diagrams

A **Venn diagram** is a diagrammatic method of picturing sets. The universal set is a rectangle. Inside the rectangle is one or more circles, each of which represents a set. Consider the experiment of stopping students at your school and asking if this is their first semester and whether they have a job. Let E be the event the student is in his/her first semester at your school and let F be the event the student has a job.

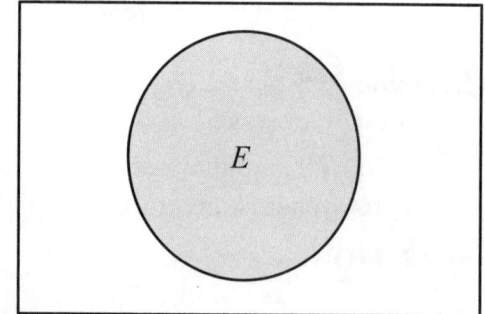

The shaded portion of the Venn diagram to the right shows the set E.

The shaded portion of the Venn diagram below shows E^C, the part of the universal set that is not in E.

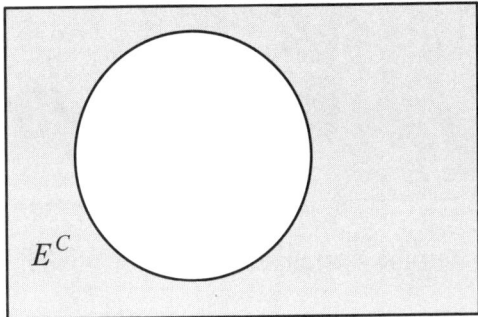

The event, E^C, is the event the student is not in his/her first semester at your school.

The shaded portion of the Venn diagram below represents the parts that E and F have in common, or the **intersection of E and F**, that is, $E \cap F$.

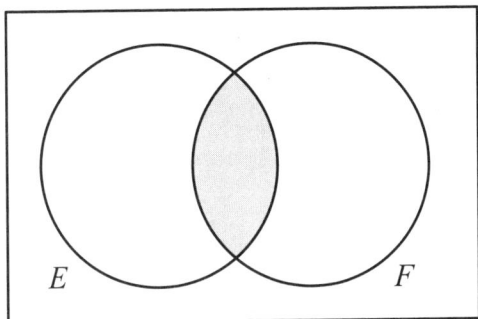

The event, $E \cap F$, is the event the student both is in his/her first semester at your school and has a job.

In the next Venn diagram, the shaded portion represents the part of E that is not in F, that is, $E \cap F^C$.

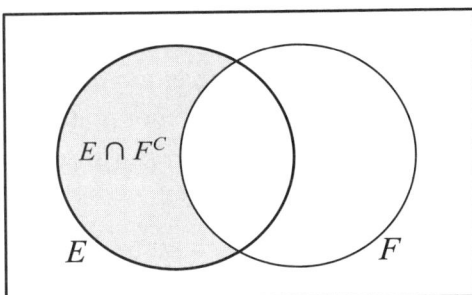

The event, $E \cap F^C$, is the event the student is in his/her first semester at your school but does not have a job.

The shaded portion of the Venn diagram on the following page represents the part that is in E or F or both, or the **union of E and F**, that is $E \cup F$.

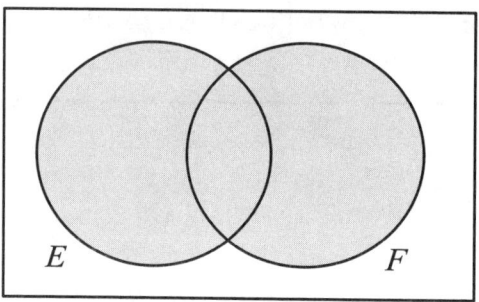

The event, $E \cup F$, is the event the student either is in his/her first semester at your school, has a job, or both.

Let E be the event that you drove your car to school today and let F be the event you do not own a car. What is the intersection, $E \cap F$? If you do not own a car, then you cannot have driven it to school today. There is nothing in the intersection, and $E \cap F$ is the empty set. We use the symbol $\emptyset$ to denote the empty set.

Since, $E \cap F$ is $\emptyset$, $P(E \cap F) = 0$. We say that E and F are **mutually exclusive**, meaning that the intersection is $\emptyset$. If E and F are mutually exclusive, then there will be no overlap between the two sets in the Venn diagram.

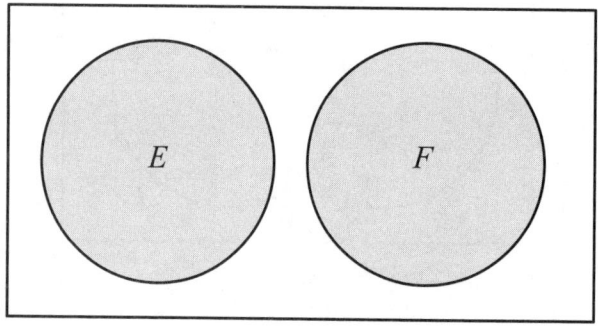

If E and F are mutually exclusive, then
$$P(E \cup F) = P(E) + P(F).$$
Let E be the event you drove your car to school today and let F be the event you own a motorcycle. What is the intersection, $E \cap F$? $E \cap F$ is the event you drove your car to school today and you own a motorcycle. Certainly, $E \cap F$ is not the empty set. So when we add $P(E)$ and $P(F)$, we've included $P(E \cap F)$ twice. Thus, if E and F are not mutually exclusive,
$$P(E \cup F) = P(E) + P(F) - P(E \cap F).$$

Example 6: Let E and F be events for which $P(E) = 0.3$, $P(F^C) = 0.6$ and $P(E \cup F) = 0.7$. What is $P(E \cap F)$? What can you say about E and F?

SOLUTION
$$P(F) = 1 - P(F^C) = 1 - 0.6 = 0.4.$$
$$\begin{aligned} P(E \cap F) &= P(E) + P(F) - P(E \cup F) \\ &= 0.3 + 0.4 - 0.7 \\ &= 0 \end{aligned}$$

Since $P(E \cap F) = 0$, E and F are mutually exclusive.

Basic Probability

Example 7: In a certain city, 45% of all families own only one car and 30% own exactly two cars. One family is selected at random. Let A be the event that the selected family owns only one car and let B be the event that the selected family owns exactly two cars.

   a. Describe the event A^C. Find $P(A^C)$.
   b. Describe the event $A \cap B$. Find $P(A \cap B)$.
   c. Describe the event $A \cup B$. Find $P(A \cup B)$.

SOLUTION

   a. A^C is the event a family does not own only one car. By the law of complements, $P(A^C) = 1 - P(A) = 1 - 0.45 = 0.55$.
   b. $A \cap B$ is the event a family owns exactly one car AND owns exactly two cars. Since this is not possible, $A \cap B$ is $\emptyset$. So, $P(A \cap B) = P(\emptyset) = 0$.
   c. $A \cup B$ is the event a family owns one car or two cars. Since a family cannot be in both categories, $P(A \cap B) = 0$, and so $P(A \cup B) = P(A) + P(B) = 0.45 + 0.30 = 0.75$.

Example 8: A survey of employees in The Company revealed that 300 people subscribe to *Newsweek*, 200 subscribe to *Time*, and 50 subscribe to both. How many people subscribe to at least one of these magazines?

SOLUTION

Let N be the event a person subscribes to *Newsweek*, and T be the event a person subscribes to *Time*. We use the notation "$n(\ )$" to indicate the number of elements in a set; for example, $n(N)$ indicates the number of people in the set N. So,

$$\text{Number of people who subscribe to at least one of these magazines}$$
$$= n \text{ (people who subscribe to at least one of these magazines)}$$
$$= n(N \cup T) = n(N) + n(T) - n(N \cap T)$$
$$= 300 + 200 - 50$$
$$= 450.$$

Example 9: Suppose $P(S) + P(T) = P(S \cup T)$. What can you say about S and T?

SOLUTION

We know that $P(S \cup T) = P(S) + P(T) - P(S \cap T)$.
$$P(S) + P(T) = P(S) + P(T) - P(S \cap T)$$
$$0 = -P(S \cap T)$$
$$0 = P(S \cap T)$$

In other words, $S \cap T$ is the null set, or S and T are mutually exclusive.

Example 10: Give a set theoretic expression that describes the shaded portion of the Venn diagram.

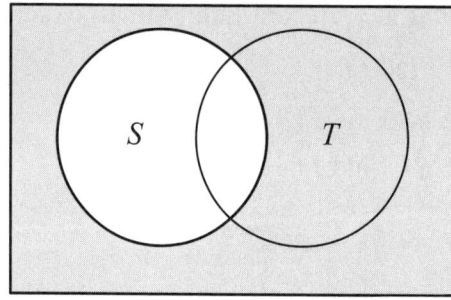

SOLUTION

The shaded portion is everything except S. That is, the colored portion is S^C.

Example 11: Give a set theoretic expression that describes the shaded portion of the Venn diagram.

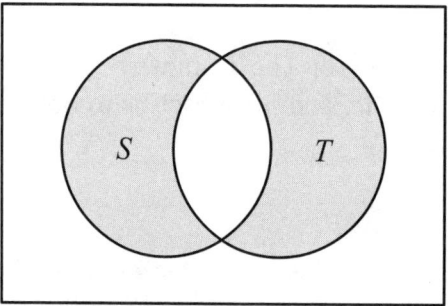

SOLUTION

The shaded portion is everything in S and in T *except* what is in the intersection $S \cap T$. That is, the shaded portion represents $(S \cap T^C) \cup (S^C \cap T)$.

Example 12: Give a set theoretic expression that describes the shaded portion of the Venn diagram.

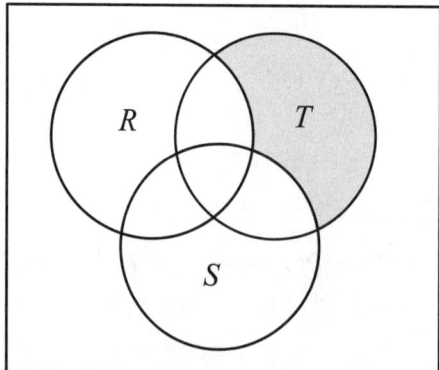

SOLUTION

The shaded portion is everything in T, but not in R nor in S. That is, the shaded portion is $T \cap R^C \cap S^C$.

Basic Probability

Example 13: Give a set theoretic expression that describes the shaded portion of the Venn diagram.

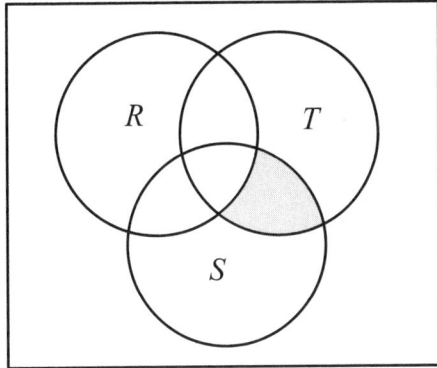

SOLUTION

The shaded portion is everything in S intersect T, but not in R. That is, the shaded portion is $S \cap T \cap R^C$.

DeMorgan's Laws are two useful properties that relate unions, intersections and complements. The first law says that if E and F are two sets,

$$E^C \cap F^C = (E \cup F)^C.$$

Let's verify the first DeMorgan's Law using Venn Diagrams.

The **union** $E \cup F$ is everything in either E or F or both. The union $E \cup F$ is the shaded portion below.

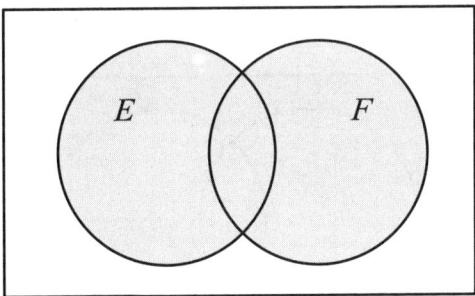

Venn Diagram 1

The **complement** of the union, $(E \cup F)^C$ is everything outside of $E \cup F$. The complement, $(E \cup F)^C$ is the shaded portion below.

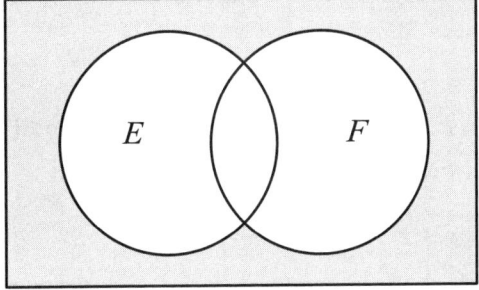

Venn Diagram 2

E^C is everything outside of E and is the shaded portion below.

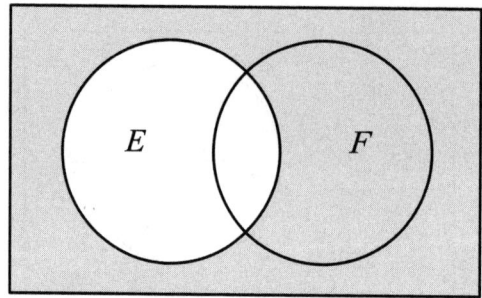

Venn Diagram 3

F^C is everything outside of F and is the shaded portion below.

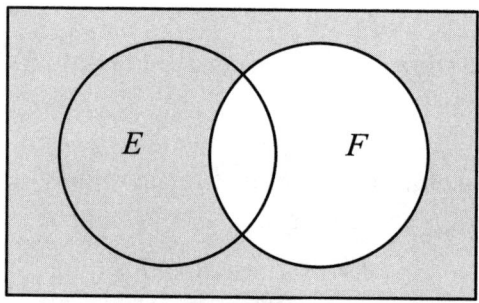

Venn Diagram 4

$E^C \cap F^C$ consists of the parts that Venn diagrams 3 and 4 have in common, that is $E^C \cap F^C$ is the shaded portion below.

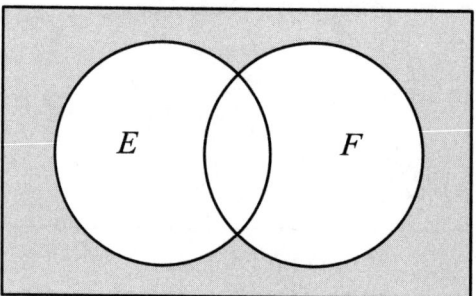

Venn Diagram 5

But Venn Diagram 5 and Venn Diagram 2 are exactly the same! Thus, we have verified the first DeMorgan's Law:
$$E^C \cap F^C = (E \cup F)^C.$$
$P(E^C \cap F^C) = P((E \cup F)^C)$ Apply the property of complements.
$\qquad\qquad\quad\; = 1 - P(E \cup F)$

Therefore, we have
$$P(E^C \cup F^C) = P((E \cap F)^C) = 1 - P(E \cap F).$$

Basic Probability

The second DeMorgan's Law says that $E^C \cup F^C = (E \cap F)^C$. Using the property of complements as above, we obtain
$$P(E^C \cup F^C) = P((E \cap F)^C) = 1 - P(E \cap F).$$

Example 14: Use DeMorgan's Laws to simplify $S^C \cup (S \cap T)^C$.

SOLUTION
$$\begin{aligned}S^C \cup (S \cap T)^C &= S^C \cup (S^C \cup T^C)\\ &= S^C \cup S^C \cup T^C\\ &= (S^C \cup S^C) \cup T^C\\ &= S^C \cup T^C\\ &= (S \cap T)^C.\end{aligned}$$

Example 15: Suppose $P(A) = 0.2$, $P(B) = 0.4$, and $P(A \cap B) = 0.15$.
  a. Find $P(A \cup B)$.
  b. Find $P(A^C)$.
  c. Find $P(A^C \cap B^C)$.
  d. Find $P(A^C \cup B^C)$.

SOLUTION
  a. $P(A \cup B) = P(A) + P(B) - P(A \cap B) = 0.2 + 0.4 - 0.15 = 0.45$.
  b. $P(A^C) = 1 - P(A) = 1 - 0.2 = 0.8$.
  c. By DeMorgan's Laws, $A^C \cap B^C = (A \cup B)^C$, so
  $$P(A^C \cap B^C) = P((A \cup B)^C) = 1 - P(A \cup B) = 1 - 0.45 = 0.55.$$
  d. Again, by DeMorgan's Laws, $A^C \cup B^C = (A \cap B)^C$, so
  $$P(A^C \cup B^C) = P((A \cap B)^C) = 1 - P(A \cap B) = 1 - 0.15 = 0.85.$$

Example 16: In a certain community, 40% of the houses need termite work, 55% need a new roof, and 35% need termite work and a new roof. A house is selected at random. Let T be the event a house needs termite work and let R be the event the house needs a new roof.
  a. Find the probability that the house does not need termite work.
  b. Find the probability the house needs either termite work or a new roof.
  c. Describe the event $T \cap R^C$. Find $P(T \cap R^C)$.
  d. Describe the event $T^C \cup R^C$. Find $P(T^C \cup R^C)$.

SOLUTION
  a. We need to find $P(T^C)$.
  $$P(T^C) = 1 - P(T) = 1 - 0.40 = 0.60$$
  b. We need to find $P(T \cup R)$.
  $$\begin{aligned}P(T \cup R) &= P(T) + P(R) - P(T \cap R)\\ &= 0.40 + 0.55 - 0.35\\ &= 0.60.\end{aligned}$$

c. $T \cap R^C$ is the event a house needs termite work but not a new roof.
$$P(T \cap R^C) = P(T) - P(T \cap R)$$
$$= 0.40 - 0.35$$
$$= 0.05$$

d. $T^C \cup R^C$ is the event a house does not need termite work or does not need a new roof. By DeMorgan's Laws, $T^C \cup R^C = (T \cap R)^C$, so
$$P(T^C \cup R^C) = P((T \cap R)^C)$$
$$= 1 - P(T \cap R)$$
$$= 1 - 0.35$$
$$= 0.65.$$

Exercises

1. An experiment consists of selecting a number at random from the set, {1, 2, 3, 4, 5, 6, 7, 8, 9}.

   a. What is the probability that the number selected is even?

   b. What is the probability that the number selected is 7 or even?

   c. What is the probability the number selected is 7 AND even?

   SOLUTION

   a. $\dfrac{4}{9}$ b. $\dfrac{5}{9}$ c. 0

2. The modern American roulette wheel has 38 slots, which are labeled with 36 numbers evenly divided between red and black, plus two green numbers 0 and 00. What is the probability that the ball lands on a green number?

   SOLUTION

   $\dfrac{1}{19}$

3. There are approximately 2.75 million telephone numbers in Arizona, of which 25% are unpublished or unlisted numbers. How many published telephone numbers are there in Arizona?

   SOLUTION

   Approximately 2.1 million

4. The table to the right shows how many schools college freshmen had applied to.

   a. What is the probability that a randomly selected freshman applied to at most two colleges?

   b. What is the probability that a randomly selected freshman applied to four or more colleges?

Number of Colleges Applied to	Percent
1	32%
2	14%
3	15%
4	11%
5 or more	28%

Basic Probability

SOLUTION

a. 0.46   b. 0.11 + 0.28 = 0.39

5. Information about the number of rooms in U.S. housing units is given in the following table.

Number of Rooms	Probability
1	0.008
2	0.012
3	0.094
4	0.190
5	0.222
6	0.202

Find the probability that a randomly selected housing unit has more than six rooms.

SOLUTION

$1 - (0.008 + 0.020 + 0.114 + 0.304 + 0.526 + 0.728) = 0.272$

6. An experiment consists of selecting a vehicle at random from a college parking lot and noting its color and make. Let E be the event "the vehicle is white," F be the event "the vehicle is a Toyota," G be the event "the vehicle is a blue Ford," and H be the event "the vehicle is a red Honda." (Assume no two-colored vehicles.)

a. Which of the following pairs of events are mutually exclusive?

   (i) E and F (ii) E and G
   (iii) F and G (iv) E and H
   (v) F and H (vi) G and H
   (vii) E^C and G (viii) F^C and H^C

b. Describe each of the following events.

   (i) $E \cap F$ (ii) $E \cup F$
   (iii) E^C (iv) F^C
   (v) G^C (vi) H^C
   (vii) $E \cup G$ (viii) $E \cap G$
   (ix) $E \cap H$ (x) $E \cup H$
   (xi) $G \cap H$ (xii) $E^C \cap F^C$
   (xiii) $E^C \cup G^C$

SOLUTION

a. (i) E and F are not mutually exclusive; the vehicle is a white Toyota.
   (ii) E and G are mutually exclusive; a vehicle cannot be both white and blue.
   (iii) F and G are mutually exclusive; a vehicle cannot be both a Ford and a Toyota.
   (iv) E and H are mutually exclusive; a vehicle cannot be both white and red.

(v) F and H are mutually exclusive; a vehicle cannot be both a Honda and a Toyota.

(vi) G and H are mutually exclusive; a vehicle cannot be both a Ford and a Honda.

(vii) E^C and G are not mutually exclusive; E^C includes all vehicles that are any color except white.

(viii) F^C and H^C are not mutually exclusive; F^C is all vehicles that are not Toyotas and H^C is all vehicles that are not red Hondas — there are vehicles that are in both F^C and H^C.

b. (i) $E \cap F$ includes all vehicles that are both white and Toyotas.

(ii) $E \cup F$ includes all vehicles that are white, or Toyotas, or both.

(iii) E^C includes all vehicles that are any color except white.

(iv) F^C includes all vehicles that are not Toyotas.

(v) G^C includes all vehicles that are not blue Fords.

(vi) H^C includes all vehicles that are not red Hondas.

(vii) $E \cup G$ includes all vehicles that are white or that are blue Fords.

(viii) $E \cap G$ is the null set, since blue Fords cannot be white.

(ix) $E \cap H$ is the null set, since red Hondas cannot be white.

(x) $E \cup H$ includes all vehicles that are white or that are red Hondas.

(xi) $G \cap H$ is the null set since a vehicle cannot be both a Ford and a Honda.

(xii) E^C includes all vehicles that are not white. F^C includes all vehicles that are not Toyotas. So $E^C \cap F^C$ includes all vehicles that are neither white nor Toyotas.

(xiii) E^C includes all vehicles that are not white. G^C includes all vehicles that are not blue Fords. So $E^C \cup G^C$ includes all vehicles.

7. A survey of 100 investors in stocks and bonds revealed that 80 investors owned stocks and 70 owned bonds. How many investors owned both stocks and bonds?

SOLUTION

50

8. Let $S = \{1, 2, 3, 4\}$, $E = \{1\}$ and $F = \{2, 3\}$. Are $E \cup F$ and $E^C \cap F^C$ mutually exclusive?

SOLUTION

Yes, $E \cup F$ and $E^C \cap F^C$ are mutually exclusive.

9. *Consumer Reports* (Jan. 2003) found that in 27% of drug ads or other promotions sent to physicians, risks were omitted, minimized or obscured; and in 18% of these ads false or misleading superiority claims were made. Suppose that in 12% of these ads both problems occurred. In how many of the ads did at least one of the problems occur?

SOLUTION

33%

Basic Probability

10. Consider the Venn diagram given below.

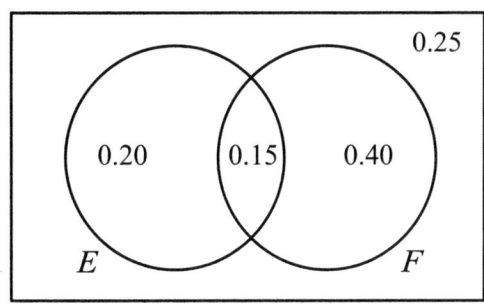

a. Find $P(E)$.
b. Find $P(F^C)$.
c. Find $P(E \cap F^C)$.
d. Find $P(E^C \cap F^C)$.

SOLUTION

a. 0.35
b. 0.45
c. 0.20
d. 0.25

11. Draw a 3-circle Venn diagram and shade the portion that corresponds to the set $(R \cup S \cup T)^C$.

SOLUTION

$R \cup S \cup T$ is everything inside the three circles,

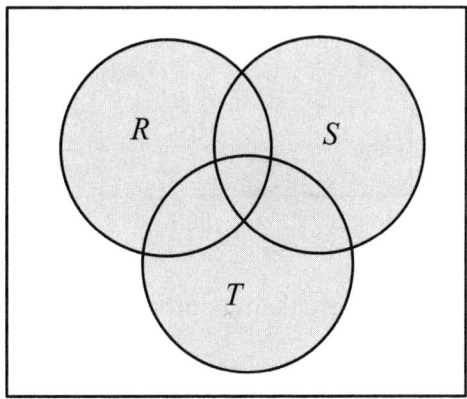

so $(R \cup S \cup T)^C$ is everything outside the three circles.

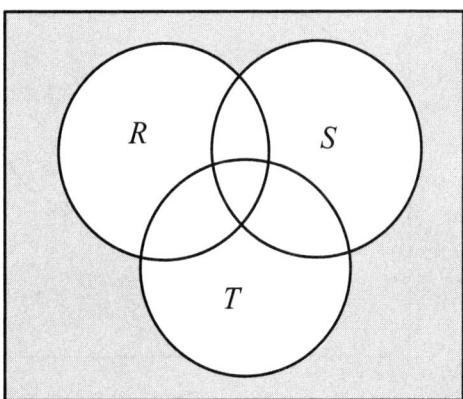

12. Draw a 3-circle Venn diagram and shade the portion that corresponds to the set $R \cap (S \cup T)$.

SOLUTION

$S \cup T$ is everything in S or T or both;

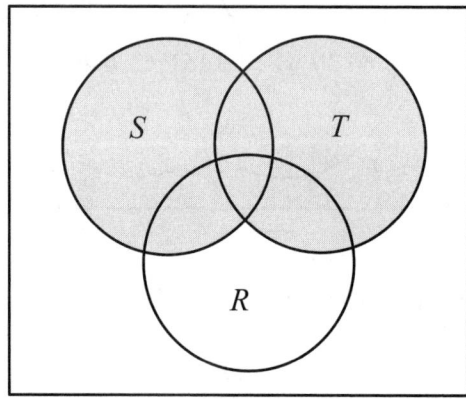

$R \cap (S \cup T)$ is the part of the above that is also in R.

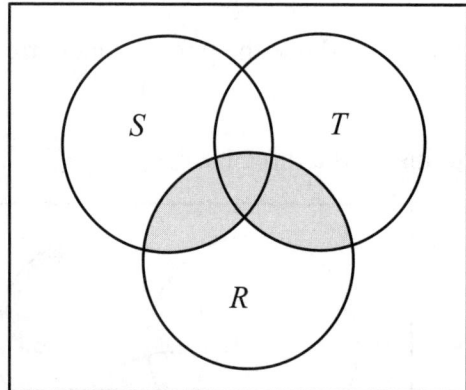

13. Draw a 3-circle Venn diagram and shade the portion that corresponds to the set $R \cap (S \cap T^C)$.

SOLUTION

$S \cap T^C$ is everything in S but not in T;

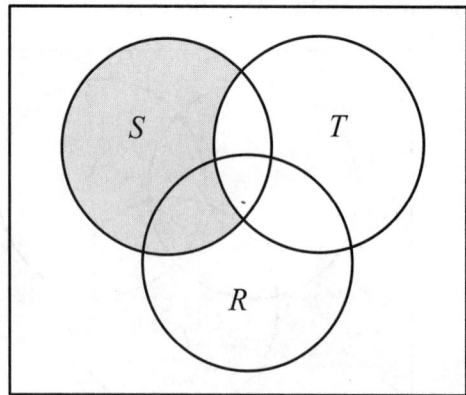

$R \cap (S \cap T^C)$ is the part of $(S \cap T^C)$ that is also in R.

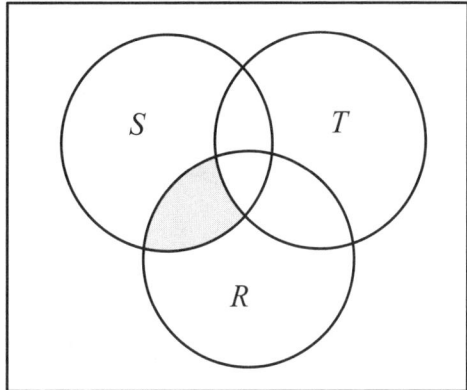

14. Let E and F be events for which $P(E) = 0.6$, $P(F) = 0.5$ and $P(E \cap F) = 0.4$.
    a. Find $P(E \cup F)$
    b. Find $P(E \cap F^C)$

SOLUTION

a. $P(E \cup F) = P(E) + P(F) - P(E \cap F)$
    $= 0.6 + 0.5 - 0.4$
    $= 0.7$.

b. $P(E \cap F^C) = P(E) - P(E \cap F) = 0.6 - 0.4 = 0.2$.

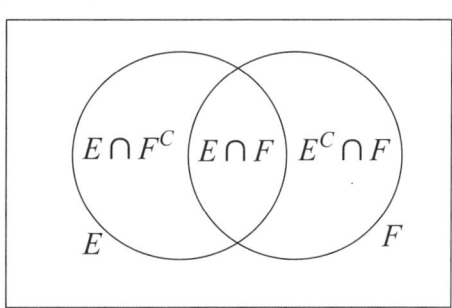

15. A farmer has 41 pigs. He notices that every fat pig is greedy. Thirty pigs are greedy, 25 are greedy and healthy, 23 are fat and greedy, and 20 are fat, healthy and greedy. Also, one pig is neither greedy nor healthy.
    a. How many pigs are healthy but not greedy?
    b. How many pigs are neither fat nor healthy?
    c. How many pigs are healthy and greedy but not fat?

SOLUTION

Let F be the event a pig is fat, H be the event a pig is healthy, and G be the event a pig is greedy.

Twenty pigs are fat, healthy, and greedy; so the number in the event $F \cap G \cap H$ is 20.

Since all fat pigs are greedy, $n(F \cap G^C \cap H^C) = 0$; for the same reason, $n(F \cap G^C \cap H) = 0$. Since all fat pigs are greedy, the pig who is neither greedy nor healthy is also not fat; thus $n((F \cup G \cup H)^C) = 1$.

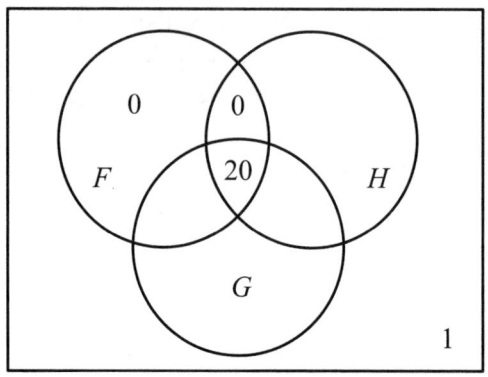

Since $n(F \cap G) = 23$ and $n(F \cap G \cap H) = 20$, $n(F \cap G \cap H^C) = 23 - 20 = 3$.
Since $n(G \cap H) = 25$ and $n(G \cap H \cap F) = 20$, $n(G \cap H \cap F^C) = 25 - 20 = 5$.

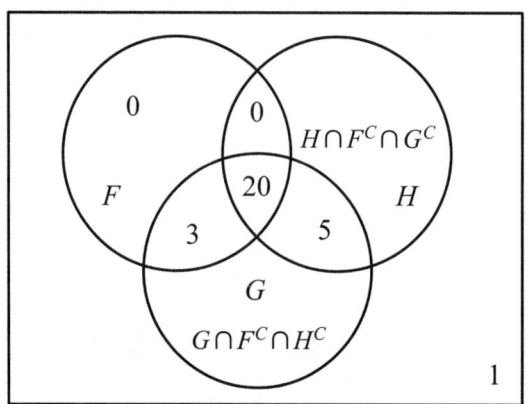

Since $n(G) = 30$, $n(G \cap F^C \cap H^C) = 30 - (3 + 20 + 5) = 2$.
Since there are 41 pigs total, $n(H \cap F^C \cap G^C) = 41 - (3 + 20 + 5 + 2 + 1) = 10$.

a. $n(H \cap G^C) = n(H \cap G^C \cap F^C) = 10$.
b. $n(F^C \cap H^C) = n(G \cap F^C \cap H^C) + n(G^C \cap F^C \cap H^C) = 2 + 1 = 3$.
c. $n(H \cap G \cap F^C) = 5$.

16. In a certain manufacturing process, the probability of a type A defect is 0.12, the probability of a type B defect is 0.22, and the probability of having both types of defects is 0.03. Find the probability of neither type of defect.

SOLUTION

0.69

17. Use DeMorgan's Laws to simplify $F \cup (E \cap F)^C$.

SOLUTION

The entire sample space, S.

18. Verify the Second DeMorgan's Law.

SOLUTION

Refer to pages 41–42 for the verification of DeMorgan's First Law.

Summation Notation

Definitions

Summation notation is a simple way to express sums that are long. For example, if we wanted to express the sum of the even integers from 2 to 100 we could easily represent this as $\sum_{i=1}^{50} 2 \cdot i$ instead of $2 + 4 + 6 + 8 + 10 + \cdots + 92 + 94 + 96 + 98 + 100$. Before we do some examples, we need to understand the various parts of summation notation, $\sum_{i=1}^{n} a_i$. This sum is read "the sum as i goes from 1 to n of a_i" where i is called the **index of summation**. The Greek letter sigma Σ is called the **summation operator**. The symbol a_i (which is a function of i) represents the **expression** that is being summed. The index of summation starts the sum at any integer value of i. The **upper limit** n of the index tells you where to end the summation. Thus,

$$\sum_{i=1}^{4} a_i = a_1 + a_2 + a_3 + a_4.$$

In summary, the **index of summation**, i, tells you where to start the sum; the **upper limit**, n, tells you where to end the sum; and a_i is what you are going to sum.

Example 1: Use summation notation to represent the sum $1 + 3 + 5 + 7 + 9 + 11$.

SOLUTION

This is the sum of six odd integers. Notice that each odd integer is one less than the next even integer. An even integer can be represented by $2 \cdot i$ where i is any integer. Therefore, an odd integer can be represented as $2 \cdot i - 1$. Since there are six integers in the sum, i will go from 1 to 6. We write this sum as

$$\sum_{i=1}^{6} (2 \cdot i - 1).$$

Example 2: Use summation notation to represent the sum $3 + 9 + 27 + 81$.

SOLUTION

We can express the sum as $3^1 + 3^2 + 3^3 + 3^4$. Notice the base of the each exponential term is 3 and the exponent increases by one in each successive term. We write this sum as

$$\sum_{i=1}^{4} 3^i.$$

As illustrated in the next example, the index of summation, i, does not have to start at 1. Also, a variable other than i can be use to represent the index of summation.

Example 3: Use summation notation to represent the sum

$$\frac{2}{3} + \frac{3}{4} + \frac{4}{5} + \frac{5}{6}.$$

Use *j* for the index of summation.

SOLUTION

Notice that the denominator of each term is one more than the numerator.

$$\sum_{j=2}^{5} \frac{j}{j+1}.$$

Example 4: Use summation notation to represent the sum $-3 + 5 - 7 + 9 - 11 + 13 - 15$. Use *k* for the index of summation.

SOLUTION

Notice that there is a sign change for each successive term, and the absolute value of each term is an odd integer starting at three. Recall that the odd integers can be represented by $2 \cdot k - 1$. Since the first odd integer in the above sum is three, *k* must start at 2. Since the signs of successive terms alternate, and the first term is negative, each term must contain a factor of $(-1)^{k-1}$. This factor will alternate in sign depending on whether $k - 1$ is even or odd. If $k - 1$ is even, then $(-1)^{k-1}$ will be positive. If $k - 1$ is odd, then $(-1)^{k-1}$ will be negative. So the sum can be written as

$$\sum_{k=2}^{8} (-1)^{k-1} \cdot (2 \cdot k - 1).$$

Example 5: Evaluate $\sum_{k=3}^{6} (2k^2 - 1)$.

SOLUTION

$$\sum_{k=3}^{6} (2 \cdot k^2 - 1) = (2 \cdot 3^2 - 1) + (2 \cdot 4^2 - 1) + (2 \cdot 5^2 - 1) + (2 \cdot 6^2 - 1)$$
$$= (2 \cdot 9 - 1) + (2 \cdot 16 - 1) + (2 \cdot 25 - 1) + (2 \cdot 36 - 1)$$
$$= 17 + 31 + 49 + 71$$
$$= 168$$

Example 6: Evaluate $\sum_{i=1}^{4} (-1)^i \cdot \ln e^{2 \cdot i}$.

SOLUTION

$$\sum_{i=1}^{4} (-1)^i \cdot \ln e^{2 \cdot i} = (-1)^1 \cdot \ln e^{2 \cdot 1} + (-1)^2 \cdot \ln e^{2 \cdot 2} + (-1)^3 \cdot \ln e^{2 \cdot 3} + (-1)^4 \cdot \ln e^{2 \cdot 4}$$
$$= -\ln e^2 + \ln e^4 - \ln e^6 + \ln e^8$$
$$= -2 + 4 - 6 + 8$$
$$= 4$$

Example 7: Evaluate $\sum_{k=2}^{5} 10$.

SOLUTION

$$\sum_{k=2}^{5} 10 = 10+10+10+10 = 40$$

Properties

Suppose that a_i and b_i represent two different expressions that we want to sum from $i = 1$ to n. We can express this in summation notation:

$$\sum_{i=1}^{n} a_i \quad \text{and} \quad \sum_{i=1}^{n} b_i.$$

The summation of the sum of or difference between two expressions is stated in Property 1.

1. $\sum_{i=1}^{n}(a_i \pm b_i) = \sum_{i=1}^{n} a_i \pm \sum_{i=1}^{n} b_i.$

For Property 2 we introduce, r, a constant that can be multiplied by any expression of a sum. From the distributive property, r can be "factored out" from the summation.

2. $\sum_{i=1}^{n} r \cdot a_i = r \cdot \sum_{i=1}^{n} a_i.$

A single sum can be broken up into two shorter sums; this leads us to Property 3.

3. $\sum_{i=1}^{n} a_i = \sum_{i=1}^{m} a_i + \sum_{i=m+1}^{n} a_i$ where $1 \leq m < n.$

A sum can also be re-indexed; this leads us to Property 4.

4. $\sum_{i=k}^{n} a_i = \sum_{j=k+c}^{n+c} a_{j-c}$ or $\sum_{i=k}^{n} a_i = \sum_{j=k-c}^{n-c} a_{j+c}$ where k and c are constants greater than zero.

This seems complicated, but it really is not. Let us examine what is happening. If n is increased by a value of c, then the new index of summation must be the original value added to c; and the variable, i, of the expression being summed must be decreased by c. The opposite also applies. If n is decreased by a value of c, then the new index of summation must be the original value less c, and the variable, i, of the expression being summed must be increased by c.

Example 8: Express $\sum_{j=3}^{10} j^2 - \sum_{j=3}^{10} 2 \cdot j$ as a single sum with k as the index of summation.

SOLUTION

From Property 1: $\sum_{k=3}^{10} (k^2 - 2 \cdot k).$

Example 9: Express $\sum_{i=1}^{5} \frac{2i+1}{3}$ as two sums. (Use Property 2 if possible.)

SOLUTION

From Property 1: $\sum_{i=1}^{5} \frac{2 \cdot i}{3} + \sum_{i=1}^{5} \frac{1}{3}$

From Property 2: $\frac{2}{3} \cdot \sum_{i=1}^{5} i + \sum_{i=1}^{5} \frac{1}{3}$

Example 10: Is $\sum_{k=2}^{100}(2 \cdot k^2 + 3 \cdot k - 5)$ equivalent to $2 \cdot \sum_{j=2}^{100} j^2 + 3 \cdot \sum_{j=2}^{100} j - \sum_{j=2}^{100} 5$?

SOLUTION

Yes, by applying Properties 1 and 2. Also, recall that changing the index of summation does not change the value of the sum.

Example 11: Express $\sum_{i=1}^{20} 3 \cdot i$ as two different sums with the first sum ending at $i = 7$.

SOLUTION

From Property 3: $\sum_{i=1}^{20} 3 \cdot i = \sum_{i=1}^{7} 3 \cdot i + \sum_{i=8}^{20} 3 \cdot i$

Example 12: Show that $\sum_{j=3}^{8}(j-2) = \sum_{j=3}^{6}(j-2) + \sum_{j=7}^{8}(j-2)$.

SOLUTION

$$\sum_{j=3}^{8}(j-2) = (3-2)+(4-2)+(5-2)+(6-2)+(7-2)+(8-2)$$

$$= [(3-2)+(4-2)+(5-2)+(6-2)] + [(7-2)+(8-2)]$$

$$= \sum_{j=3}^{6}(j-2) + \sum_{j=7}^{8}(j-2)$$

Note that the sum in the first set of brackets corresponds to $\sum_{j=3}^{6}(j-2)$ and the sum in the second set of brackets corresponds to $\sum_{j=7}^{8}(j-2)$.

Example 13: Re-index the sum $\sum_{i=5}^{10} 2 \cdot i$ with the new index of summation starting at $j = 2$.

SOLUTION

Since the index is decreasing by three $(5 - 3 = 2)$, the upper limit of the index must also decrease by three $(10 - 3 = 7)$. The variable in the expression must be increased by three $(i = j + 3)$.

$$\sum_{j=2}^{7} 2 \cdot (j+3)$$

Summation Notation

Example 14: Verify that the answer in Example 13 is correct.

SOLUTION

We need to show that $\sum_{i=5}^{10} 2 \cdot i = \sum_{j=2}^{7} 2 \cdot (j+3)$.

$$\sum_{i=5}^{10} 2 \cdot i = 2 \cdot 5 + 2 \cdot 6 + 2 \cdot 7 + 2 \cdot 8 + 2 \cdot 9 + 2 \cdot 10$$
$$= 2 \cdot (5+6+7+8+9+10)$$
$$= 2 \cdot (45)$$
$$= 90$$

$$\sum_{j=2}^{7} 2 \cdot (j+3) = 2 \cdot (2+3) + 2 \cdot (3+3) + 2 \cdot (4+3) + 2 \cdot (5+3) + 2 \cdot (6+3) + 2 \cdot (7+3)$$
$$= 2 \cdot 5 + 2 \cdot 6 + 2 \cdot 7 + 2 \cdot 8 + 2 \cdot 9 + 2 \cdot 10$$
$$= 2 \cdot (5+6+7+8+9+10)$$
$$= 2 \cdot (45)$$
$$= 90$$

Example 15: Re-index the sum $\sum_{i=1}^{3}(i^2+1)$ with the new index of summation starting at $k = 5$.

SOLUTION

Since the index is increasing by four ($1 + 4 = 5$), the upper limit of the index must also increase by four ($3 + 4 = 7$). The variable in the expression must decrease by four ($i = k - 4$).

$$\sum_{i=1}^{3}(i^2+1) = \sum_{k=5}^{7}\left[(k-4)^2 + 1\right]$$

Example 16: Verify that the answer in Example 15 is correct.

SOLUTION

We need to show that $\sum_{i=1}^{3}(i^2+1) = \sum_{k=5}^{7}\left[(k-4)^2 + 1\right]$.

$$\sum_{i=1}^{3}(i^2+1) = (1^2+1) + (2^2+1) + (3^2+1)$$
$$= 2 + 5 + 10$$
$$= 17$$

$$\sum_{k=5}^{7}[(k-4)^2+1] = [(5-4)^2+1] + [(6-4)^2+1] + [(7-4)^2+1]$$
$$= (1^2+1) + (2^2+1) + (3^2+1)$$
$$= 2 + 5 + 10$$
$$= 17$$

So far we have done all of our calculations by hand. If the sums are very large, it would not be practical to do the computations by hand. For example, we would not want to compute $\sum_{i=1}^{100}(i^2 - 2\cdot i + 1)$ by hand. It would be more practical to use a spreadsheet such as *Excel* to compute sums with more than ten terms. In the following examples, *Excel* is used for the computations.

Example 17: Use Excel to compute $\sum_{j=3}^{25}(3\cdot j - 20)^2$.

SOLUTION

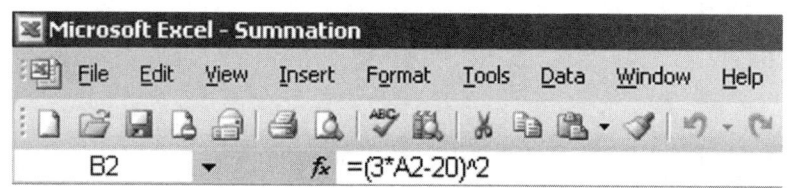

j	$(3 \cdot j - 20)^2$
3	121
4	64
5	25
6	4
7	1
8	16
9	49
10	100
11	169
12	256
13	361
14	484
15	625
16	784
17	961
18	1156
19	1369
20	1600
21	1849
22	2116
23	2401
24	2704
25	3025
Sum =	20,240

Example 18: Re-index the sum in Example 17 with an upper limit of 30 rather 25. Use *Excel* to compute the new sum. (Note that your answer should be the same as in Example 17.)

SOLUTION

$$\sum_{k=8}^{30}(3\cdot(k-5)-20)^2 = \sum_{k=8}^{30}(3\cdot k - 15 - 20)^2$$
$$= \sum_{k=8}^{30}(3\cdot k - 35)^2$$

Summation Notation

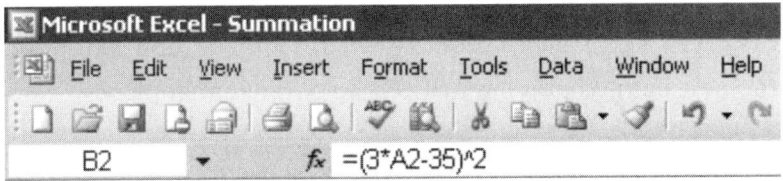

k	$(3 \cdot k - 35)^2$
8	121
9	64
10	25
11	4
12	1
13	16
14	49
15	100
16	169
17	256
18	361
19	484
20	625
21	784
22	961
23	1156
24	1369
25	1600
26	1849
27	2116
28	2401
29	2704
30	3025
Sum = 20,240	

Example 19: Use *Excel* to compute $\sum_{i=10}^{100} \frac{i-1}{i+1}$. **Show the last 10 terms of the sum,** and give your answer to three decimal places.

SOLUTION

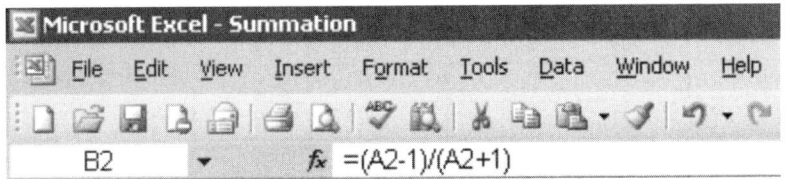

91	0.978261
92	0.978495
93	0.978723
94	0.978947
95	0.979167
96	0.979381
97	0.979592
98	0.979798
99	0.980000
100	0.980198
Sum = 86.463	

Example 20: Use *Excel* to compute $\sum_{i=12}^{35} \frac{2 \cdot i}{i^2 - 30}$. Then re-index the sum with an upper limit of 30 rather than 35, and evaluate the new sum. Give your answers to two decimal places.

SOLUTION

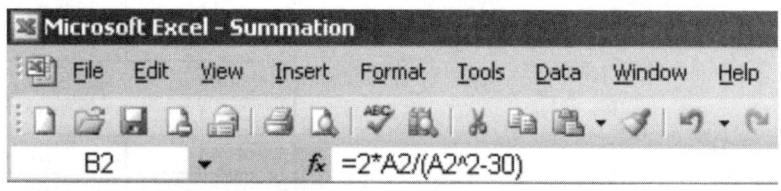

$$\sum_{i=12}^{35} \frac{2 \cdot i}{i^2 - 30} \qquad \sum_{j=7}^{30} \frac{2 \cdot (j+5)}{(j+5)^2 - 30}$$

i	$2 \cdot i / (i^2 - 30)$	j	$2 \cdot (j+5)/[(j+5)^2 - 30]$
12	0.210526316	7	0.210526316
13	0.187050360	8	0.187050360
14	0.168674699	9	0.168674699
15	0.153846154	10	0.153846154
16	0.141592920	11	0.141592920
17	0.131274131	12	0.131274131
18	0.122448980	13	0.122448980
19	0.114803625	14	0.114803625
20	0.108108108	15	0.108108108
21	0.102189781	16	0.102189781
22	0.096916300	17	0.096916300
23	0.092184369	18	0.092184369
24	0.087912088	19	0.087912088
25	0.084033613	20	0.084033613
26	0.080495356	21	0.080495356
27	0.077253219	22	0.077253219
28	0.074270557	23	0.074270557
29	0.071516646	24	0.071516646
30	0.068965517	25	0.068965517
31	0.066595059	26	0.066595059
32	0.064386318	27	0.064386318
33	0.062322946	28	0.062322946
34	0.060390764	29	0.060390764
35	0.058577406	30	0.058577406
	Sum = 2.49		**Sum = 2.49**

Exercises

1. Use summation notation to represent the sum $2 + 4 + 6 + 8 + 10 + 12 + 14$.

SOLUTION

$$\sum_{i=1}^{7} 2 \cdot i$$

2. Use summation notation, with k as the index of summation, to represent the sum $3 - 5 + 7 - 9 + 11 - 13 + 15 - 17 + 19$.

SOLUTION

$$\sum_{k=2}^{10} (-1)^k \cdot (2k-1)$$

3. Use summation notation, with j as the index of summation, to represent the sum

$$-\frac{\sqrt{2}}{3} + \frac{\sqrt{3}}{4} - \frac{\sqrt{4}}{5} + \cdots - \frac{\sqrt{20}}{21} + \frac{\sqrt{21}}{22}.$$

SOLUTION

$$\sum_{j=2}^{21} \left((-1)^{j-1} \cdot \frac{\sqrt{j}}{j+1} \right)$$

4. Use summation notation to represent the sum $100 + 100 + 100 + 100 + 100$.

SOLUTION

$$\sum_{i=1}^{5} 100$$

5. Evaluate $\sum_{i=1}^{5} (i-1)^i$.

SOLUTION

$$\sum_{i=1}^{5} (i-1)^i = 1114$$

6. Evaluate $\sum_{k=3}^{7} \frac{2 \cdot k}{k-1}$ and express your answer as an improper fraction.

SOLUTION

$$\sum_{k=3}^{7} \frac{2 \cdot k}{k-1} = \frac{129}{10}$$

7. Evaluate $\sum_{i=2}^{5} \sqrt[i]{\log(i/2)}$ and round each term to five decimal places.

SOLUTION

$$\sum_{i=2}^{5} \sqrt[i]{\log(i/2)} = 2.13292$$

8. Re-index the sum in Exercise 6 with the index of summation starting at $i = 1$. Verify that your answer is correct.

SOLUTION

$$\sum_{i=1}^{5} \frac{2 \cdot i + 4}{i+1} = \frac{129}{10}$$

9. If $\sum_{i=2}^{6}(3 \cdot i - 2) = \sum_{j=6}^{n} a_j$, find n and a_j. Verify that your answer is correct.

SOLUTION

$n = 10$ and $a_j = 3 \cdot j - 14$

$$\sum_{i=2}^{6}(3 \cdot i - 2) = \sum_{j=6}^{10}(3 \cdot j - 14) = 50$$

10. Evaluate $\sum_{i=6}^{300} \frac{2 \cdot i}{i-5}$ using *Excel*. Give your answer to two decimal places.

SOLUTION

$$\sum_{i=6}^{300} \frac{2 \cdot i}{i-5} = 652.66$$

11. Evaluate $\sum_{k=1}^{25}(2 \cdot k - 10) - \sum_{k=5}^{20}(\sqrt{k} + 1)$ using *Excel*. Give your answer to two decimal places.

SOLUTION

$$\sum_{k=1}^{25}(2 \cdot k - 10) - \sum_{k=5}^{20}(\sqrt{k} + 1) = 328.48$$

12. Re-index the sum in Exercise 11 with the first index starting at $j = 5$ and the second index starting at $i = 1$. Verify your answer using *Excel*.

SOLUTION

$$\sum_{j=5}^{29}(2 \cdot j - 18) - \sum_{i=1}^{16}(\sqrt{i+4} + 1) = 328.48$$

13. Re-index $\sum_{i=12}^{60}(3 \cdot i^2 + 5)$ with an upper limit of 55 rather than 60. Use *Excel* to verify your answer.

SOLUTION

$$\sum_{j=7}^{55}(3 \cdot (j+5)^2 + 5) = 220{,}157$$

14. Evaluate $\sum_{i=5}^{50} \frac{2-i}{i-4}$ using *Excel*. Give your answer to three decimal places.

SOLUTION

$$\sum_{i=5}^{50} \frac{2-i}{i-4} = -54.833$$

15. True or False? (You do not need to use *Excel* to answer this question.)

a. $\sum_{k=2}^{100} \ln(e+k) = \sum_{j=20}^{118} \ln(e+j+18)$

b. $\sum_{i=100}^{500} (i^2 + 2 \cdot i + 3) = \sum_{j=50}^{450} (j^2 + 102 \cdot j + 2603)$

SOLUTION
   a. False
   b. True

Expected Value

A **random variable** X is a rule that assigns a numerical value to every outcome in the sample space. We use the capital letter X to indicate the variable; we use a lower case letter x to indicate values that the random variable X takes.

The **expected value** of a random variable X is the theoretical average we would obtain if we repeated the experiment many, many times. We use the notation $E(X)$ for expected value.

If each outcome is equally likely, the expected value is the arithmetic average of the values X takes. Consider the experiment of tossing one die. Let X be the number of spots on the top face. X can take the values 1, 2, 3, 4, 5, or 6. All are equally likely, so the expected value of X is the average of the six possible outcomes. In other words,

$$E(X) = \frac{1+2+3+4+5+6}{6} = \frac{21}{6} = \frac{7}{2} = 3.5 \ .$$

Of course, on any one toss of the die, we couldn't get 3.5 spots. But if we threw the die many, many times, the average would be approximately 3.5.

Now consider another random variable X, where X is the number of boys in a 2-child family. X can take the values 0, 1, 2. There is one way for X to equal 0, namely both children are girls, or GG. There is one way for X to equal 2, namely both children are boys, or BB. There are two ways for X to equal 1, namely BG and GB. So the probability distribution of the random variable X is as follows:

x	$P(X=x)$
0	1/4 = 0.25
1	1/2 = 0.50
2	1/4 = 0.25

Since the probabilities of the three outcomes are not all the same, we couldn't just add the numbers 0, 1, 2 and divide by 3 to find the expected value of X. X will take the value 1 about two times out of every four and will take the value 0 or 2 about one time each out of every four. So we "weight" the outcomes according to each probability. That is,

$$E(X) = 0 \cdot 0.25 + 1 \cdot 0.5 + 2 \cdot 0.25 = 1.$$

In general, then, to find the expected value of a random variable X, we use the formula

$$E(X) = \sum_{\text{all } x} x \cdot P(X = x).$$

Example 1: In the coming year, the probability is 0.3 that the price of a certain house will go up 20%, the probability is 0.5 that the price will go up 10%, the probability is 0.1 that the price will remain the same, and the probability is 0.1 that the price will go down 10%. Find the expected percentage increase.

SOLUTION

Let X be the percentage increase of the price of the house.

x	$P(X=x)$	$x \cdot P(X=x)$
20	0.3	6
10	0.5	5
0	0.1	0
−10	0.1	−1

$$E(X) = \sum_{\text{all } x} x \cdot P(X = x) = 10.$$

The price of the house is expected to rise 10% in the coming year.

Example 2: In a lottery, there are five first prizes worth $100 each, ten second prizes worth $50 each, and 20 third prizes worth $25 each. Suppose 10,000 tickets are sold. Tickets cost $2 each. Find the expected value of one ticket.

Expected Value

SOLUTION

Let X be the net amount you win or lose on a ticket.

x	$P(X=x)$	$x \cdot P(X=x)$
98	5/10,000 = 0.0005	0.049
48	10/10,000 = 0.001	0.048
23	20/10,000 = 0.002	0.046
-2	9965/10,000 = 0.9965	-1.993

$$E(X) = \sum_{\text{all } x} x \cdot P(X=x) = -\$1.85.$$

The expected value of a ticket is a loss of $1.85.

Example 3: Suppose an urn contains 4 white balls and 7 red balls. Choose two balls, without replacement (that is, you choose the first ball and do not put it back before you choose the second ball).
a. Write all the possible outcomes of this experiment.
b. If X is the random variable which gives the number of red balls you get, give the probability distribution of X.
c. What is $E(X)$?

SOLUTION

a. Let W stand for the event "the ball you draw is white" and let R stand for the event "the ball you draw is red." There are four possible outcomes: WW, WR, RW, and RR.
b. If X is the number of red balls drawn, X can take the value 0, 1, or 2. The probability that X takes the value 0 is the same as the probability that you get two white balls. On the first draw, there are 11 balls, of which 4 are white; if you got a white ball on the first draw, there are 10 balls left for the second draw, of which only 3 are white. Thus,

$$P(X=0) = P(WW) = \frac{4}{11} \cdot \frac{3}{10} = \frac{6}{55}.$$

Similarly,

$$P(X=1) = P(WR \text{ or } RW) = \frac{4}{11} \cdot \frac{7}{10} + \frac{7}{11} \cdot \frac{4}{10} = \frac{28}{55}$$

$$P(X=2) = P(RR) = \frac{7}{11} \cdot \frac{6}{10} = \frac{21}{55}.$$

So the probability distribution of X is given as follows:

x	$P(X=x)$
0	6/55
1	28/55
2	21/55

Note that the sum of the probabilities is exactly 1.

c. $E(X) = \sum_{\text{all } x} x \cdot P(X = x) = 0 \cdot \frac{6}{55} + 1 \cdot \frac{28}{55} + 2 \cdot \frac{21}{55} = \frac{70}{55} = 1.27.$

Example 4: A standard deck of cards has 52 cards, 13 of each of the four suits. Choose three cards with replacement. Let X be the number of Diamonds in the three cards.

a. What are the values that X can take?
b. What is the probability distribution of X?
c. Find $E(X)$.

SOLUTION

a. X can take the value 0, 1, 2, or 3.

b. Let D be the event you draw a Diamond and let N be the event you draw a card that is not a Diamond.

x	Cards Needed	$P(X = x)$
0	NNN	$\frac{39}{52} \cdot \frac{39}{52} \cdot \frac{39}{52} = \frac{27}{64}$
1	NND or NDN or DNN	$3 \cdot \frac{39}{52} \cdot \frac{39}{52} \cdot \frac{13}{52} = \frac{27}{64}$
2	NDD or DND or DDN	$3 \cdot \frac{39}{52} \cdot \frac{13}{52} \cdot \frac{13}{52} = \frac{9}{64}$
3	DDD	$\frac{13}{52} \cdot \frac{13}{52} \cdot \frac{13}{52} = \frac{1}{64}$

Note that the sum of the probabilities is exactly 1.

c. $E(X) = \sum_{\text{all } x} x \cdot P(X = x) = 0 \cdot \frac{27}{64} + 1 \cdot \frac{27}{64} + 2 \cdot \frac{9}{64} + 3 \cdot \frac{1}{64} = \frac{48}{64} = \frac{3}{4}.$

Example 5: In the game of "Lucky", three dice are rolled. It costs $1 to play. For each 6 that comes up, the player wins $1. Find the expected winnings for one game.

SOLUTION

Let X be the net amount won on each roll. Note that the probability of a 6 on each die is 1/6 and the probability of any other number showing up on each die is 5/6.

$$E(X) = \sum_{\text{all } x} x \cdot P(X = x) = -\frac{108}{216} = -0.50.$$

We expect to lose $0.50 on a game.

Number of 6's rolled	x	$P(X=x)$	$x \cdot P(X=x)$
0	-1	$\frac{5}{6} \cdot \frac{5}{6} \cdot \frac{5}{6} = \frac{125}{216}$	$-\frac{125}{216}$
1	0	$3 \cdot \frac{5}{6} \cdot \frac{5}{6} \cdot \frac{1}{6} = \frac{75}{216}$	0
2	1	$3 \cdot \frac{5}{6} \cdot \frac{1}{6} \cdot \frac{1}{6} = \frac{15}{216}$	$\frac{15}{216}$
3	2	$\frac{1}{6} \cdot \frac{1}{6} \cdot \frac{1}{6} = \frac{1}{216}$	$\frac{2}{216}$
Total		1	$-\frac{108}{216}$

Exercises

1. If a first offender is convicted of a certain crime, the probability is 0.1 that the judge will sentence him to three years in jail, the probability is 0.2 that the sentence will be only one year, and the probability is 0.7 that he will be placed on probation. What is the expected length of time a first offender will be sentenced to jail?

SOLUTION

0.5 year

2. When Kaylee goes out in the morning on her usual 4-mile route, she walks slowly (20-minute miles) for 5% of the time, walks quickly (15-minute miles) for 25% of the time, and runs the remainder of the time. If she burns 84 calories per mile walking slowly, 95 calories per mile walking quickly and 125 calories per mile running, how many calories does she burn doing her usual route?

SOLUTION

About 462 calories.

3. A bag contains 6 pieces of bubble gum, 10 caramels, and 6 toffees. You reach in the bag and, without looking and without replacement, choose two candies. Let X be the number of caramels you get. Find $E(X)$.

SOLUTION

$E(X) = \frac{10}{11}$

4. An urn contains 7 white balls and 10 black balls. Choose 2 balls, without replacement. Let X be the number of black balls that you get. Find $E(X)$.

SOLUTION

$$E(X) = \frac{20}{17}$$

5. A publisher introduces a new weekly magazine that sells for $5.95. The company's marketing department estimates that the sales X, in thousands, will be approximated by the probability distribution given below.

x	15	20	25	30	35
$P(X=x)$	0.30	0.35	0.20	0.10	0.05

What is $E(X)$?

SOLUTION

$E(X) = 21.25$

6. A company pays its new employees $51,000 if they have an MBA and $33,600 if they have a BA in Business. The probability that a new employee will have an MBA is 0.32 and the probability that a new employee will have a BA in Business is 0.68. What is the mean starting salary for these new employees?

SOLUTION

$E(X) = \$39,168$

7. Let X be the number of people in a household in the United States. For 2000, the U.S. Census reported that X has the probability distribution given below.

x	1	2	3	4	5
$P(X=x)$	0.258	0.326	0.165	0.142	0.108

What is the average size of a household in the United States?

SOLUTION

$E(X) = 2.513$

8. Hurricanes are classified by categories, where a category 1 storm is the weakest and a category 5 storm is the strongest. Let X be the category of a hurricane which has hit the U.S. mainland. Then according to *USA Today* Weather Almanac, X has the probability distribution given below.

X	1	2	3	4	5
$P(X=x)$	0.372	0.229	0.288	0.098	0.013

What is $E(X)$?

SOLUTION

$E(X) = 2.151$

Conditional Probability

Definitions

If we are given information about a situation, and then want the probability of something else, that is called **conditional probability**. For example, suppose you know that a family with two children has at least one girl. Then you are restricting the possible outcomes to the set {*GG, GB, BG*} only. Knowing that the family has at least one girl, what is the probability the other child is a girl as well? Of the three outcomes in the restricted set, only one of them corresponds to two girls. So, if we know that one child is a girl, the probability the other child is also a girl is 1/3.

The additional piece of information is the part we are given. The "probability of *E* given that *F* has occurred" is called **the conditional probability of *E* given *F*** and is written $P(E|F)$. To calculate the conditional probability we use the formula

$$P(E|F) = \frac{P(E \cap F)}{P(F)}.$$

If we multiply both sides of this equation by $P(F)$, we have

$$P(E \cap F) = P(E|F) \cdot P(F).$$

Since the conditional probability of *F* given *E* can be calculated using the formula

$$P(F|E) = \frac{P(F \cap E)}{P(E)},$$

we can multiply both sides of this equation by $P(E)$, and obtain

$$P(E \cap F) = P(F|E) \cdot P(E).$$

Previously, using unions and intersections, we found that

$$P(E \cap F) = P(E) + P(F) - P(E \cup F).$$

Thus, there are three formulas for the probability of the intersection of *E* and *F*, $P(E \cap F)$. Use whichever one of the three is most convenient in a particular situation.

If the outcomes are equally likely, then we can compute the conditional probability by counting the number of outcomes in each event, that is,

$$P(E|F) = \frac{\frac{n(E \cap F)}{n(S)}}{\frac{n(F)}{n(S)}} = \frac{n(E \cap F)}{n(F)}.$$

Example 1: Following a major earthquake, a team of engineers visited a city to assess whether or not buildings constructed under the new earthquake safety guidelines had fared better than older buildings. Every building in a 6-square block downtown area was inspected, and the following results were obtained.

	New Construction	Old Construction	Total
No significant damage	18	75	93
Moderate damage	26	57	83
Severe damage	24	62	86
Total	68	194	262

One of these buildings is selected at random.

a. Find the probability that the selected building has no significant damage.
b. Find the probability that the selected building has no significant damage, given that it is of new construction.
c. Find the probability that the selected building has severe damage.
d. Find the probability that the selected building has severe damage, given that it is of new construction.

SOLUTION

a. P(no significant damage)
$$= \frac{\text{number with no significant damage}}{\text{total number of buildings}}$$
$$= \frac{93}{262}$$
$$\cong 0.355.$$

b. P(no significant damage | new construction)
$$= \frac{\text{number with no significant damage and new construction}}{\text{total number of buildings of new construction}}$$
$$= \frac{18}{68}$$
$$\cong 0.2647.$$

c. P(severe damage)
$$= \frac{\text{number with severe damage}}{\text{total number of buildings}}$$
$$= \frac{86}{262}$$
$$\cong 0.3282.$$

d. P(severe damage | new construction)
$$= \frac{\text{number of new construction and severe damage}}{\text{total number of new construction}}$$
$$= \frac{24}{68}$$
$$\cong 0.3529.$$

Example 2: It is known that 1% of pet turtles smell bad. There are two types of pet turtles, green and brown. The owner of a pet store discovers that 90% of his smelly turtles are green, while only 50% of his other (non-smelly) turtles are green.

a. Find the overall percentage of green turtles.
b. If a single turtle is selected at random, find the probability that it is smelly, if we know it is green.

Conditional Probability

SOLUTION

Define the following events:

S is the event a turtle smells bad.

NS is the event a turtle is non-smelly.

G is the event a turtle is green.

B is the event a turtle is brown.

What we are given is the following:

$P(S) = 0.01$

$P(G|S) = 0.90$

$P(G|NS) = 0.50$

This information is organized in the tree diagram to the right.

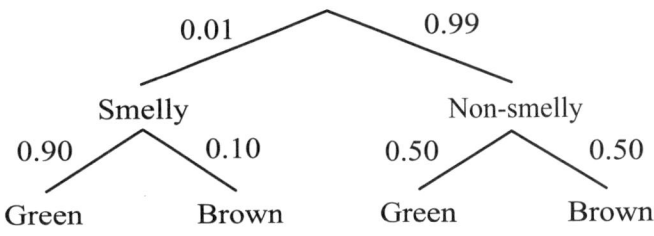

a. $P(G) = P((G \cap S) \cup (G \cap NS))$
    $= P(G \cap S) + P(G \cap NS)$
    $= P(G|S) \cdot P(S) + P(G|NS) \cdot P(NS)$
    $= (0.90) \cdot (0.01) + (0.50) \cdot (0.99)$
    $= 0.504$

So 50.4% of his turtles are green.

b. $P(S|G) = \dfrac{P(S \cap G)}{P(G)} = \dfrac{(0.90) \cdot (0.01)}{0.504} \cong 0.018$

Independent Events

Sometimes knowing an additional piece of information changes the probability of an event; sometimes that additional information does not change the probability of the event.

For example, consider the experiment of tossing a fair coin twice and recording whether we get heads or tails on each toss. There are four possible outcomes: {HH, HT, TH, TT}. Let E be the event that the second toss is a head. There are two outcomes in the event E, namely {HH, TH}; the probability of E is $P(E) = 2/4 = 1/2$.

Now suppose you know the first toss was a head; call F the event the first toss was a head. We are restricted to the two outcomes {HH, HT}. Knowing the first toss was a head, the probability the second toss is a head is $P(E|F) = 1/2$. Knowing the first toss was a head does not change the probability of the second toss being a head. We say that the two events, "head on the first toss" and "head on the second toss" are **independent** because knowing that one happened does not change the probability that the other happened.

Mathematically, we can determine that two events E and F are independent if

$$P(E|F) = P(E).$$

If E and F are independent, then the probability of their intersection reduces to

$$P(E \cap F) = P(E|F) \cdot P(F) = P(E) \cdot P(F).$$

Let's again consider the experiment of tossing a fair coin twice and recording whether we get heads or tails on each toss. Let G be the event that both tosses are heads; the probability of G is $P(G) = 1/4$. Now, suppose you know that F happened, that is, that the first toss was a head; then the possible outcomes are restricted to {HH, HT}. If we know the first toss was a head, then the probability the second toss was also a head is $P(G|F) = 1/2$. Knowing the additional piece of information, the first toss was a head, changed the probability that we got two heads on two tosses. So G and F are **not independent.**

It is important to distinguish between independent events and mutually exclusive events. E and F are independent if $P(E \cap F) = P(E) \cdot P(F)$. Two events, E and F, are mutually exclusive if $P(E \cap F) = 0$.

In the first example above, the outcomes in E, the event that the second toss is a head, are {HH, TH}; the outcomes in F, the event that the first toss is a head, are {HH, HT}. We saw above that E and F are independent, but they are not mutually exclusive, since the intersection $E \cap F$ is the outcome {HH}.

In the second example above, the outcome in G, the event that both tosses are heads, is {HH}. In this case, G and F are not mutually exclusive nor are they independent.

Example 3: A doctor studies the known cancer patients in a certain town. The probability that a randomly chosen resident has cancer is found to be 0.001. It is found that 30% of the town works for Titanic Compounds. The probability that an employee of Titanic has cancer is 0.003. Are the events "has cancer" and "works for Titanic" independent events? Are they mutually exclusive?

SOLUTION

Let T be the event a resident works for Titanic Compounds.

Let C be the event a resident has cancer.

$$P(C|T) = \frac{P(C \cap T)}{P(T)} = \frac{0.003}{0.3} = 0.01.$$

$$P(C) = 0.001.$$

Since these two are not equal, the events C and T are not independent.
The events are not mutually exclusive, since we know there are people who work for Titanic Compounds who do have cancer.

Example 4: Fifty-five percent of children attending kindergarten have not had a chicken pox vaccine. In a class of 18 students, what is the probability that at least one child has not had a chicken pox vaccine?

SOLUTION

Unless there are twins in the class, we can assume that whether one child in the class has had the vaccine is independent of whether any other child in the class has had the vaccine. Let X be the random variable that gives the number of children in the class who have not had the vaccine. X can take any integer value from 0 to 18. Then

$$\begin{aligned} P(X \geq 1) &= 1 - P(X = 0) \\ &= 1 - (1 - 0.55)^{18} \\ &= 1 - (0.45)^{18} \\ &= 0.99999943. \end{aligned}$$

Example 5: The Houston Rockets' new player, Hou Yao, has a field goal percentage of 0.586 (as of 12/20/02). Assuming independence, what is the probability that Yao will make both of his next two field goal attempts? What is the probability he will make neither? What is the probability he will make exactly one?

SOLUTION

Since we assume independence, the probability that Yao makes any field goal attempt is 0.586. Let $S1$ be the event the first attempt is successful, $S2$ be the event the second attempt is successful, $N1$ be the event the first attempt is not successful, and $N2$ be the event the second attempt is not successful).
Then the probability he makes both of his next two attempts is

$$P(S1 \cap S2) = P(S1) \cdot P(S2) = (0.586) \cdot (0.586) \cong 0.3434.$$

The probability he makes neither of his next two attempts is

$$P(N1 \cap N2) = P(N1) \cdot P(N2) = (1-0.586)^2 = (0.414)^2 \cong 0.1714.$$

The probability he makes exactly one of his next two attempts is

$$P(\text{one is successful}) = 1 - (P(\text{both are successful}) + P(\text{neither is successful}))$$
$$\cong 1 - (0.3434 + 0.1714)$$
$$= 0.4852.$$

Example 6. Sixty percent of the residents of a community have, at one time, attended the local community college. Thirty-eight percent of the residents favor a tax override for the community college and 35% of the residents are people who have both attended the community college and support the tax override.

a. What is the probability that a person chosen at random supports the tax override, given that the person has attended the community college?

b. What is the probability that a person chosen at random attended the community college, if we know that person supports the tax override?

c. What is the probability that a randomly chosen person did not attend the community college and does not support the tax override?

SOLUTION

a. Let T be the event a resident supports the tax override and let C be the event the person attended the local community college. We know that

$$P(C) = 0.6$$
$$P(T) = 0.38$$
$$P(T \cap C) = 0.35.$$

So $P(T|C) = \dfrac{P(T \cap C)}{P(C)} = \dfrac{0.35}{0.60} \cong 0.583$.

b. $P(C|T) = \dfrac{P(C \cap T)}{P(T)} = \dfrac{0.35}{0.38} \cong 0.921$.

c. We are looking for $P(T^C \cap C^C)$. By DeMorgan's Laws, $T^C \cap C^C = (T \cup C)^C$. Also,

$$P(T \cup C) = P(T) + P(C) - P(T \cap C)$$
$$= 0.38 + 0.6 - 0.35 = 0.63.$$

Thus,

$$P(T^C \cap C^C) = P((T \cup C)^C)$$
$$= 1 - P(T \cup C) = 1 - 0.63 = 0.37.$$

Example 7. In a carnival game, the player selects two coins from a bag containing three silver dollars and seven slugs. That is, the player selects one coin and, without replacing it, selects a second coin. How much should the player pay to play the game so that, over the long run, he breaks even?

SOLUTION

There are four possible outcomes. The probability of what you get on the second pick is not independent of the first pick, so the probability of what you get on the second pick changes.

Let D be the event that the player selects a silver dollar and let S be the event that the player selects a slug.

Outcome	Probability
DD	$\frac{3}{10} \cdot \frac{2}{9} = \frac{6}{90} = \frac{2}{30}$
DS	$\frac{3}{10} \cdot \frac{7}{9} = \frac{21}{90} = \frac{7}{30}$
SD	$\frac{7}{10} \cdot \frac{3}{9} = \frac{21}{90} = \frac{7}{30}$
SS	$\frac{7}{10} \cdot \frac{6}{9} = \frac{42}{90} = \frac{14}{30}$

Suppose it costs $\$p$ to play the game. Let X be the random variable that gives the net earnings on one play of the game.

Outcome	x	$P(X=x)$	$x \cdot P(X=x)$
DD	$2-p$	2/30	$(2-p) \cdot (2/30)$
DS, SD	$1-p$	14/30	$(1-p) \cdot (14/30)$
SS	$0-p = -p$	14/30	$(-p) \cdot (14/30)$

Thus

$$E(X) = \sum_{\text{all } x} x \cdot P(X=x) = (2-p) \cdot \left(\frac{2}{30}\right) + (1-p) \cdot \left(\frac{14}{30}\right) + (-p) \cdot \left(\frac{14}{30}\right).$$

If the player is to break even over the long run, we want $E(X) = 0$. So,

$$(2-p) \cdot \left(\frac{2}{30}\right) + (1-p) \cdot \left(\frac{14}{30}\right) + (-p) \cdot \left(\frac{14}{30}\right) = 0$$

$$4 - 2 \cdot p + 14 - 14 \cdot p - 14 \cdot p = 0$$

$$-30 \cdot p = -18$$

$$p = \frac{18}{30} = \frac{3}{5} = 0.60.$$

Therefore, to break even over the long run, the game should cost $0.60.

Exercises

1. Suppose a pair of dice is tossed. Given that a double does not occur, what is the probability that exactly one die shows a 4?

SOLUTION

$\frac{1}{3}$

2. At a recent murder trial, there were 150 people in the jury pool. Sixty of those people were eliminated because they were strongly in favor of the death penalty. Fifteen additional people were eliminated from the pool for other reasons. If one of the eliminated potential jurors is selected at random, what is the probability that person strongly favors the death penalty?

SOLUTION

$\frac{4}{5}$

3. Suppose a family has three children. Let E be the event there is at most one boy and let F be the event there is at least one of each gender.
   a. Are E and F independent?
   b. Are E and F mutually exclusive?

SOLUTION

   a. E and F are independent.
   b. E and F are not mutually exclusive.

4. A magician claims to have ESP powers. He puts down on a table three cards, numbered 1 to 3. While he is looking the other way, you move the cards around so he no longer knows which is which. The magician then tries to guess the identity of each card. He cannot guess the same number more than once and he cannot look at the cards until he has made all three guesses. Assume that the magician is guessing.
   a. What is the probability that all three guesses are correct?
   b. What is the probability that no guesses are correct?

c. What is the probability that exactly one guess is correct?

d. What is the probability that exactly two guesses are correct?

SOLUTION

a. $\dfrac{1}{6}$ b. $\dfrac{1}{6}$ c. $\dfrac{2}{3}$ d. 0

5. The New York Times reported (3/20/05) that for every violent death in Congo's war zone, there are 62 nonviolent deaths from the conflict. The following table gives the causes of these nonviolent deaths:

	Malnutrition	Respiratory disease and diarrhea	Anemia, measles, meningitis, accidents, tuberculosis	Fever	Other causes
Children 5 and under	3	5	4	11	5
Children 5 to 14	1	1	1	2	1
Women 15 and older	1	2	2	2	6
Men 15 and older	1	3	3	2	6

a. What is the probability that a randomly chosen victim who dies from a nonviolent cause is a child 5 and under?

b. Given that the victim was a child 5 and under, what is the probability the child died from fever?

SOLUTION

a. $\dfrac{14}{31}$

b. $\dfrac{11}{28}$

6. The strength of an earthquake is measured using the Richter scale, in which smaller numbers mean weaker earthquakes and larger numbers indicate stronger earthquake. (The Dec. 26, 2004 earthquake, which spawned huge tsunamis, registered 9.0.) The National Earthquake Information Center reports that worldwide there were 1,307 earthquakes in 2004 measuring at least 1 on the Richter Scale. These were distributed as follows:

Magnitude	Number of Earthquakes
8.0 to 9.9	1
7.0 to 7.9	4
6.0 to 6.9	39
5.0 to 5.9	247
4.0 to 4.9	675
3.0 to 3.9	231
2.0 to 2.9	84
1.0 to 1.9	26

a. What is the probability that an earthquake in 2004 was at least magnitude 6?

b. If you know that an earthquake in 2004 was at least magnitude 5, what is the probability that is of magnitude 6.0 to 6.9?

SOLUTION

a. 0.0337

b. 0.1340

7. The United States Census reports that in 2000, 75.1% of the U.S. population was White, 12.3% was Black, and 12.5% were Other Races. It also reports that 15.8% of all households consisted of 1 person, 32.6% consisted of 2 persons, 16.5% consisted of 3 persons, 14.2% consisted of 4 persons and 10.8% consisted of 5 or more persons. Assuming independence, what is the probability that a randomly selected person is White and lives in a 4-person household?

SOLUTION

0.1066

8. To graduate from high school, students must take a Mathematics Exam and a Written Essay. Of all students who take the exam, 36% fail the Mathematics Exam. Of those who fail the Mathematics Exam, 75% also fail the Written Essay. Of those who pass the Mathematics Exam, 20% fail the Written Essay.

a. What is the probability that a randomly chosen student fails the Written Essay?

b. Let M be the event a student fails the Mathematics Exam and W be the event that a student fails the Written Essay. Are M and W independent events? Explain.

SOLUTION

a. 0.398

b. No, since $P(W) \neq P(W|M)$.

9. Of all Internet users, 60% own exactly one computer, 25% log on for more than 30 hours per week, and 12% both own exactly one computer and log on for more than 30 hours per week.

a. If a randomly chosen Internet user owns exactly one computer, what is the probability that the he or she logs on for more than 30 hours a week?

b. If a randomly chosen Internet user logs on for more than 30 hours per week, what is the probability that the he or she owns exactly one computer?

SOLUTION

a. 0.20

b. 0.48

10. The American Lung Association reported that, in 2002, 22.1% of all American adults were smokers, 23.9% of all adult males were smokers, and 20.0% of all adult females were smokers. If 49% of all American adults are male, what is the probability that a randomly chosen adult is a male smoker?

SOLUTION

0.117

Bayes' Theorem

Suppose there is a test that detects a certain disease 93% of the time; however, 3% of the time, it indicates that a person has the disease when they do not (this is called a false positive). Suppose further that the people in a large urban area are routinely tested for the disease, and that it is estimated that 2.5% of these people actually have the disease. If the test is positive for a particular person, what is the probability the person actually has the disease?

Define the following events:

D is the event the person has the disease.

ND is the event the person does not have the disease.

T is the event the test is positive.

N is the event the test is negative.

We are looking for the probability of D, given that we know that T has occurred, that is, we want the conditional probability $P(D|T)$. By conditional probability,

$$P(D|T) = \frac{P(T \cap D)}{P(T)}.$$

The difficulty is that T can happen in two ways: the test can be positive and the person has the disease, or the test can be positive and the person does not have the disease. That is,

$$P(T) = P(T \cap D) + P(T \cap ND).$$

We also know from conditional probability that $P(T \cap D) = P(T|D) \cdot P(D)$ and that $P(T \cap ND) = P(T|ND) \cdot P(ND)$. Thus

$$P(D|T) = \frac{P(D \cap T)}{P(T)}$$

$$= \frac{P(T|D) \cdot P(D)}{P(T|D) \cdot P(D) + P(T|ND) \cdot P(ND)}.$$

Bayes' Theorem

This last expression is an illustration of Bayes' Theorem.

In order to calculate the conditional probability $P(D|T)$, we first create a tree diagram to organize the given information.

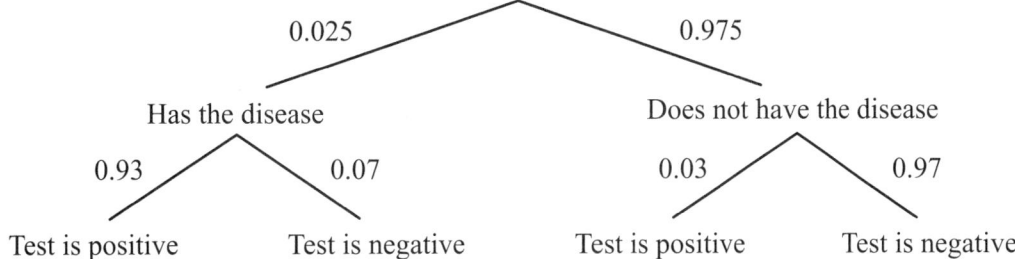

Then using Bayes' Formula we get

$$P(D|T) = \frac{P(D \cap T)}{P(T)}$$

$$= \frac{P(T|D) \cdot P(D)}{P(T|D) \cdot P(D) + P(T|ND) \cdot P(ND)}$$

$$= \frac{(0.93) \cdot (0.025)}{(0.93) \cdot (0.025) + (0.03) \cdot (0.975)}$$

$$= \frac{0.02325}{0.0525}$$

$$\cong 0.4429.$$

Example 1: It is well known that the occurrence of red-green color blindness is more prevalent in men than in women. It is believed that 22% of men have red-green color blindness while only 2% of women do. The adult population is 51% female. If you choose a person at random, what is the probability they are colorblind?

SOLUTION

A person can be colorblind in two ways: the person can be a man and be colorblind, or the person can be a woman and be colorblind. The following tree diagram organizes the information we are given:

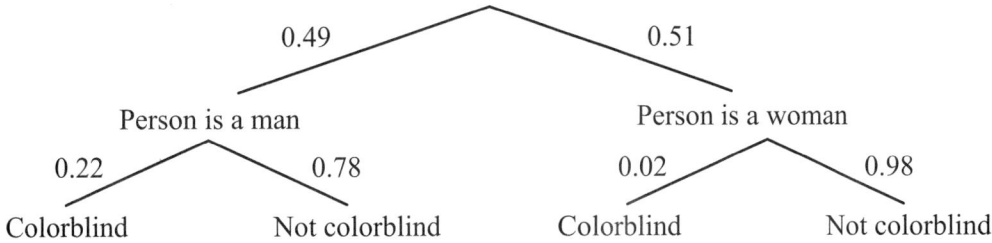

Define the following events:
  M is the event the person is a man;
  W is the event the person is a woman;
  C is the event the person is colorblind.

Since the event C is the union of two disjoint events, $C = (C \cap M) \cup (C \cap W)$, the probability of C is $P(C) = P(C \cap M) + P(C \cap W)$. From Conditional Probability, $P(C \cap M) = P(M) \cdot P(C|M)$ and $P(C \cap W) = P(W) \cdot P(C|W)$. Reading down the branches of the tree diagram, $P(C \cap M) = P(M) \cdot P(C|M) = 0.49 \cdot 0.22 = 0.1078$ and $P(C \cap W) = P(W) \cdot P(C|W) = 0.51 \cdot 0.02 = 0.0102$. Thus,

$$P(C) = P(C \cap M) + P(C \cap W) = 0.1078 + 0.0102 = 0.118.$$

Example 2: Superior Supermarket has four employees who package and weigh produce. Arturo weighs the packages accurately 98% of the time; Betsey has a 97% accuracy rate; Cliff has a 96% accuracy rate; and Dolores has a 95% accuracy rate. Arturo and Betsey each package 30% of the produce, Cliff packages 25% of the produce, and Dolores packages 15%. What is the probability that a produce package is inaccurately labeled?

SOLUTION

Inaccurately labeled produce packages could have been packaged by Arturo, Betsey, Cliff, or Dolores. The information we are given can be organized in a tree diagram:

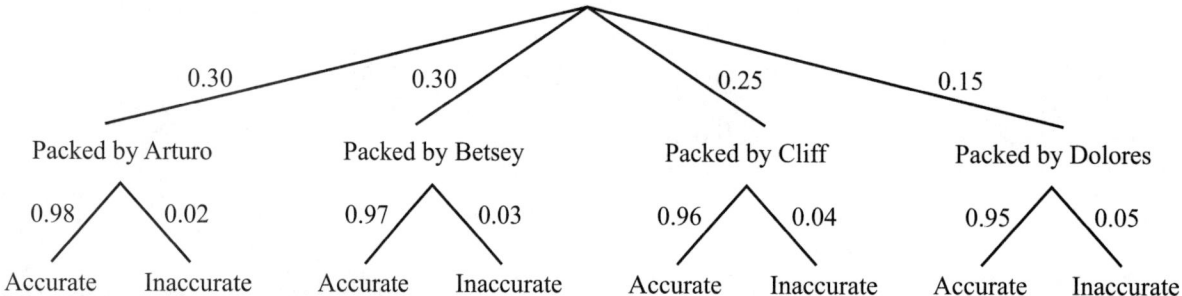

Define the following events:
 A is the event the package was packed by Arturo;
 B is the event the package was packed by Betsey;
 C is the event the package was packed by Cliff;
 D is the event the package was packed by Dolores;
 I is the event the package is inaccurately labeled.

The event I is the union of four disjoint events:

$$I = (I \cap A) \cup (I \cap B) \cup (I \cap C) \cup (I \cap D)$$

and the probability is

$$\begin{aligned} P(I) &= P(I \cap A) + P(I \cap B) + P(I \cap C) + P(I \cap D) \\ &= P(I|A) \cdot P(A) + P(I|B) \cdot P(B) + P(I|C) \cdot P(C) + P(I|D) \cdot P(D) \\ &= (0.02) \cdot (0.30) + (0.03) \cdot (0.30) + (0.04) \cdot (0.25) + (0.05) \cdot (0.15) \\ &= 0.0325 \end{aligned}$$

The probability a package is labeled inaccurately is 0.0325.

Example 3: You love the bread from EATS Bakery, one of two independent bakeries in your small town. But the older La Boulangerie is cheaper and 70% of the bread bought at an independent bakery is bought there. You, however, find that half the time the bread from La Boulangerie, while it looks the same as the bread from EATS Bakery, is too hard and has no taste.

You go to a friend's home for dinner. You like the bread served; your friend tells you she bought it at a bakery. What is the probability that your friend bought the bread at La Boulangerie?

SOLUTION

Let E be the event the bread was bought at EATS Bakery and B be the event the bread was bought at La Boulangerie. Let L be the event you like the bread and N be the event you do not like the bread. We create a tree diagram displaying the information we are given about the probabilities.

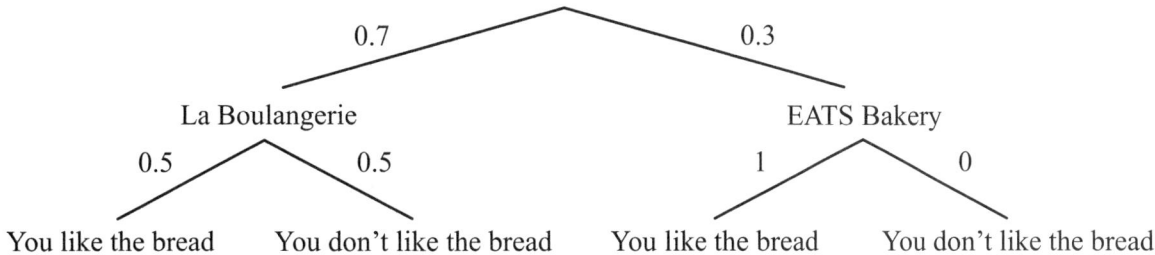

$$P(B \mid L) = \frac{P(B \cap L)}{P(L)}$$

$$= \frac{P(B \cap L)}{P(L \cap B) + P(L \cap E)}$$

$$= \frac{P(L \mid B) \cdot P(B)}{P(L \mid B) \cdot P(B) + P(L \mid E) \cdot P(E)}$$

$$= \frac{(0.5) \cdot (0.7)}{(0.5) \cdot (0.7) + (1) \cdot (0.3)}$$

$$\cong 0.5385.$$

Example 4: In a recent New York Times article, it was reported that light trucks, which include SUVs, pick-up trucks and minivans, accounted for 40% of all personal vehicles on the road in 2002. Assume the rest are cars. Of every 100,000 car accidents, 20 involve a fatality; of every 100,000 light truck accidents, 25 involve a fatality. If a fatal accident is chosen at random, what is the probability the accident involved a light truck?

SOLUTION

Let T be the event the vehicle is a light truck, C be the event the vehicle is a car, F be the event the accident involved a fatality, and NF the event the accident did not involve a fatality. We are given the following:

$P(F|C) = \dfrac{20}{100000}$

$P(F|T) = \dfrac{25}{100000}$

$P(T) = 0.4$

$P(C) = 0.6$.

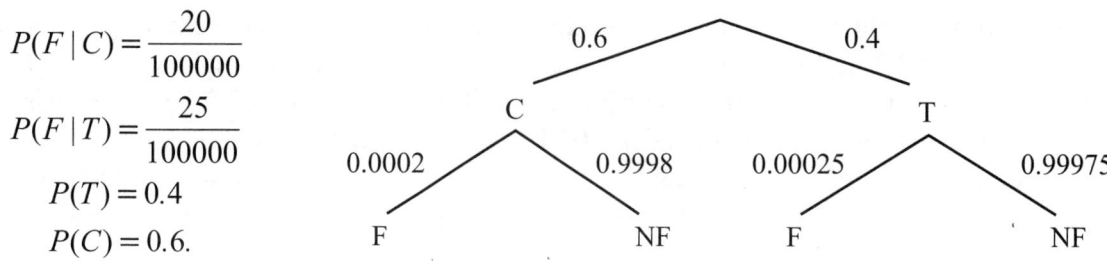

We are looking for the conditional probability that the vehicle was a light truck, given that there was a fatality.

$$P(T|F) = \frac{P(T \cap F)}{P(F)}$$

$$= \frac{P(T \cap F)}{P(F \cap T) + P(F \cap C)}$$

$$= \frac{P(F|T) \cdot P(T)}{P(F|T) \cdot P(T) + P(F|C) \cdot P(C)}$$

$$= \frac{(0.00025) \cdot (0.4)}{(0.00025) \cdot (0.4) + (0.0002) \cdot (0.6)}$$

$$\cong 0.4545.$$

Example 5: Each of three envelopes contains two bills. One envelope has two $5 bills. The second envelope has two $10 bills. The last envelope has one $5 and one $10 bill. You choose an envelope at random and remove one of the bills. It is a $5 bill. What is the probability that the other one is also a $5 bill?

SOLUTION

If the first bill removed from the envelope is a $5 bill, then it is not possible that the other one is also a $5 bill unless you chose the envelope with two $5 bills. Let $E1$ be the event that you choose the envelope with two $5 bills, $E2$ be the event that you choose the envelope with two $10 bills, $E3$ be the event that you choose the envelope with one $5 bill and one $10 bill, and $F1$ be the event that the first bill removed from the envelope is a $5 bill. We want $P(E1|F1)$. Using Bayes' Theorem,

$$P(E1|F1) = \frac{P(E1 \cap F1)}{P(F1)}$$

$$= \frac{P(E1 \cap F1)}{P(F1 \cap E1) + P(F1 \cap E2) + P(F1 \cap E3)}$$

$$= \frac{P(F1|E1) \cdot P(E1)}{P(F1|E1) \cdot P(E1) + P(F1|E2) \cdot P(E2) + P(F1|E3) \cdot P(E3)}$$

$$= \frac{(1) \cdot (1/3)}{(1) \cdot (1/3) + (0) \cdot (1/3) + (1/2) \cdot (1/3)}$$

$$= \frac{1/3}{1/2} = \frac{2}{3}.$$

Bayes' Theorem

Exercises

1. A camera is found to be defective. It is known that of all cameras of this type produced, 30% come from Factory I, 45% come from Factory II, and 25% come from Factory III. It is also known that of all cameras produced at Factory I, 1.5% are defective; of those produced at Factories II and III, 2% and 3%, respectively, are defective.

    a. What is the probability that the defective camera came from Factory I?

    b. From Factory II?

    c. From Factory III?

    SOLUTION

    a. 0.2143

    b. 0.4286

    c. 0.3571

2. A large group consists of 55% men and 45% women. Of the men, 5% are redheads, while 25% of the women are redheads. If a member of this group chosen at random is found to be a redhead, what is the probability that the person is man?

    SOLUTION

    0.1964

3. Two different suppliers, A and B, provide a manufacturer with the same part. All supplies of this part are kept in a large bin. Historically, 4% of the parts supplied by A and 10% of the parts supplied by B have been defective. A supplies three times as many parts as B. Suppose you reach into the bin, select a part, and find that it is defective. What is the probability that it was supplied by B?

    SOLUTION

    $\frac{5}{11}$

4. An auto insurance company charges younger drivers a higher premium than it does older drivers. (Younger drivers, as a group, are involved in more accidents.) The company has three age groups: Group A includes those under 25 years old, and is 18% of all policyholders. Group B includes those 25–49 years old, and is 43% of all policyholders. Group C includes those 50 years old and older. Company records show that in any given one-year period, 12% of its Group A policyholders have an accident, 3% of Group B policyholders have an accident, and 5% of Group C policyholders have an accident.

    a. What percent of the company's policyholders are expected to have an accident during the next 12 months?

    b. Suppose Ms. Y has just had a car accident. If she is one of the company's policyholders, what is the probability that she is under 25?

SOLUTION
   a. 5.4%
   b. 0.4

5. Suppose 1.5% of all individuals in a population have a certain disease. A diagnostic test for the disease is not very accurate, because 4.9% of the total population tests positive. For a person who has the disease, there is an 80% chance that the test will be positive.

   a. What is the probability that someone who doesn't have the disease will test positive?

   b. What is the probability that a person who tests positive actually has the disease?

SOLUTION
   a. 0.038
   b. 0.245

Project 2: Stock Option Pricing

Business Background

Companies issue bonds as well as stock in order to raise funds for capital expenditures. When a company sells a corporate bond it is actually borrowing money from the investor. The company pays interest on a regular basis (generally semiannually) and repays the amount borrowed on the maturity date of the bond. Specialized investment banks handle most of the trading of corporate bonds. When a company issues stock it is actually selling an ownership position in the company. Shareholder's equity is equal to the value of the assets owned by the company minus the liabilities. Each share of stock gives the holder one vote in the management of the company as well as a residual claim on the assets owned by the company. Earnings are distributed in the form of dividends on a regular basis (generally quarterly). After the initial offering, stock is sold through brokers and dealers on organized exchanges such as the New York Stock Exchange (NYSE) and the American Stock Exchange (AMEX), on regional exchanges such as the Pacific Stock Exchange (PSE) and the Chicago Stock Exchange (CSE), and in the over-the-counter (OTC) market. The market price of a stock reflects investors' expectations about the future prospects of the issuing company.

A stock option is a derivative security: its value is derived from the underlying stock. A stock option is a contract that gives the holder (buyer) of the option the right, but not the obligation, to buy or sell 100 shares of the underlying stock at a specified price on or before a specified date. The holder of an option is not obligated to exercise the right given in the contract; however, the writer (seller) of the option is obligated to honor the terms of the contract if the option is exercised. A call option gives the holder the right to buy the underlying stock, and a put option gives the holder the right to sell the underlying stock. Call options generally become more valuable as the value of the underlying stock *increases*, and put options generally become more valuable as the value of the underlying stock *decreases*. The strike (or exercise) price is the price specified in the options contract. It is the per share price at which the underlying stock can be bought or sold under the terms of the contract. The expiration date is the date on which the options contract terminates. For options that are traded on a national exchange in the United States, the expiration date is usually the third Friday of the month. An American option may be exercised prior to the expiration date, but a European option may only be exercised on the expiration date. Options are traded on organized exchanges such as the Chicago Board Options Exchange (CBOE) and in the over-the-counter market.

An American Call Option

Consider an American call option on Microsoft stock with a strike price of $27.50 that expires on July 15, 2005. The price of the option was $0.15 on May 20, 2005. If the closing price of the stock on July 15th is greater than the strike price, then the value of the call option is equal to the difference between the closing price and the strike price. Otherwise the value of the call option is equal to $0 since the investor is not obligated to exercise the right given in the contract. For example, the

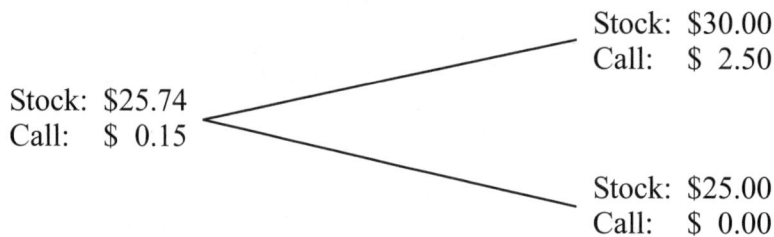

value of the option on July 15th would be equal to $2.50 if the closing price of the stock is $30.00 and $0 if the closing price of the stock is $25.00. This information is summarized in the diagram above.

Suppose that the closing price of the stock on July 15, 2005 is $30.00. Since the closing price of the stock is greater than the strike price, the call option would be exercised. The holder of the option could purchase 100 shares of Microsoft stock at the strike price of $27.50 per share and sell those same shares on the open market for $30.00 per share. The computation of the total profit (excluding transaction costs) is shown below.

$$\begin{aligned}\text{Total profit} &= [(\text{closing price} - \text{strike price}) - \text{option price}] \cdot 100 \\ &= [(30.00 - 27.50) - 0.15] \cdot 100 \\ &= 2.35 \cdot 100 \\ &= \$235.00\end{aligned}$$

The holder of the option could also purchase and hold on to the shares of Microsoft stock.

On the other hand, suppose that the closing price of the stock on July 15, 2005 is $25.00. Since the closing price of the stock is less than the strike price, the call option would not be exercised. Remember that the holder of the option is not obligated to buy the stock. The only cost is the initial price paid for the contract. The computation of the total cost (excluding transaction costs) is shown below.

$$\begin{aligned}\text{Total cost} &= \text{option price} \cdot 100 \\ &= 0.15 \cdot 100 \\ &= \$15.00\end{aligned}$$

As was stated previously, the value of a call option increases as the value of the underlying stock increases. The diagram below shows the relationship between the profit from the purchase of the call option on Microsoft stock and the value of the stock.

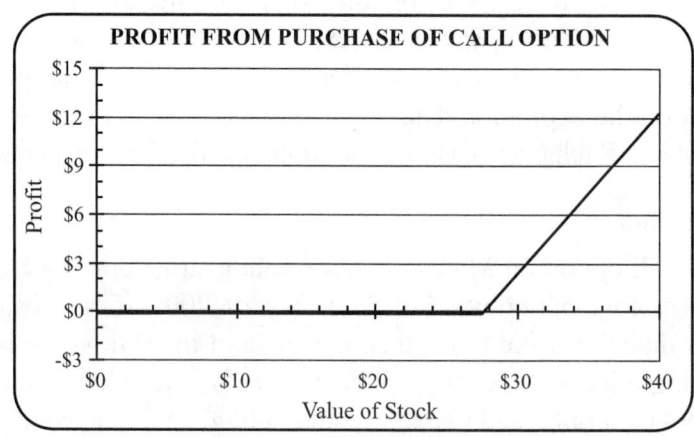

An American Put Option

Consider an American put option on Microsoft stock with a strike price of $27.50 that expires on July 15, 2005. The price of the option was $1.80 on May 20, 2005. If the closing price of the stock on July 15th is less than the strike price, then the value of the put option is equal to the difference between the strike price and the closing price. Otherwise the value of the put option is equal to $0 since the investor is not obligated to exercise the right given in the contract. For example, the value of the option on July 15th would be equal to $0 if the closing price of the stock is $30.00 and $2.50 if the closing price of the stock is $25.00. This information is summarized in the following diagram.

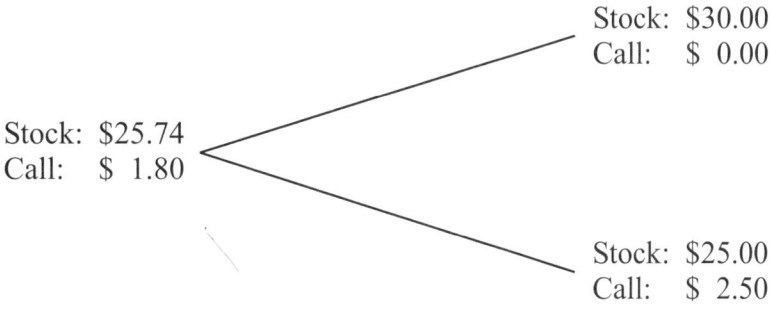

Suppose that the closing price of the stock on July 15, 2005 is $30.00. Since the closing price of the stock is greater than the strike price, the put option would not be exercised. Remember that the holder of the option is not obligated to sell the stock. The only cost is the initial price paid for the contract. The computation of the total cost (excluding transaction costs) is shown below.

$$\begin{aligned} \text{Total cost} &= \text{option price} \cdot 100 \\ &= 1.80 \cdot 100 \\ &= \$18.00 \end{aligned}$$

On the other hand, suppose that the closing price of the stock on July 15, 2005 is $25.00. Since the closing price of the stock is less than the strike price, the put option would be exercised. The holder of the option could purchase 100 shares of Microsoft stock on the open market for $25.00 per share and sell those same shares to the writer of the option for $27.50 per share. The computation of the total profit (excluding transaction costs) is shown below.

$$\begin{aligned} \text{Total profit} &= [(\text{strike price} - \text{closing price}) - \text{option price}] \cdot 100 \\ &= [(27.50 - 25.00) - 1.80] \cdot 100 \\ &= 0.70 \cdot 100 \\ &= \$70 \end{aligned}$$

The holder of the option could also sell shares of Microsoft stock that were previously purchased.

As was stated previously, the value of a put option increases as the value of the underlying stock *decreases*. The diagram given below shows the relationship between the profit from the purchase of the put option on AOL Time Warner stock and the value of the stock.

Can you think of a reason why you would want to purchase a stock option rather than purchasing the stock?

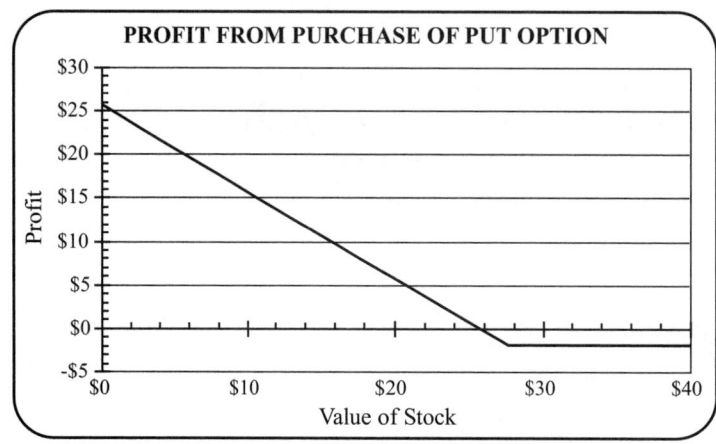

The price of an option reflects investors' expectations about the value of the option on the expiration date. More specifically, the price of the option is a function of:
1. The current stock price (C).
2. The strike price (S).
3. The time until the expiration date (t).
4. The risk free rate of interest (r_{rf}).
5. The volatility in the returns of the underlying stock.

The Black-Scholes Model is a very complicated function that is used to determine the price of call options. Since the mathematics used in the Black-Scholes Model is beyond the scope of this course, a procedure called bootstrapping will be used to find the prices of call options. A large number of samples selected with replacement from one representative sample will be used to approximate the distribution of the value of a call option on the expiration date. The present value of the mean of that distribution is then used to estimate the price of the option. There is a relationship between the prices of put and call options on the same stock with the same strike price and exercise date. Therefore, the value of the put option can be computed from the value of the call option.

Please visit `http://www.cboe.com` for additional information about stock options.

Alternative Project 2: Managing ATM Queues

Business Background

People often have to wait in line at places such as airports, banks, post offices, restaurants, supermarkets, traffic lights, etc. However, waiting lines do not always involve people. Waiting lines may also form when a bottleneck occurs on a factory assembly line, a computer's central processing unit receives instructions, a store takes delivery of new inventory, etc. Such lines are more formally called queues, and the mathematical study of queues is called queuing theory.

Queuing systems generally consist of input (such as the arrival of customers), one or more queues or lines, one or more servers, and output (such as the departure of customers). They are commonly configured in one of two ways. In the first, a separate queue is formed at each server. The customer or unit at the front of the queue is routed to the server when it becomes available. This is referred to as the standard model and is illustrated in the diagram given below.

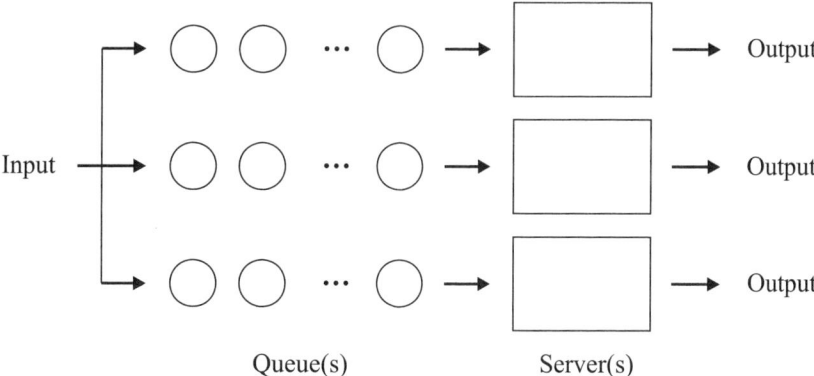

In the second, a single queue is formed. The customer or unit at the front of the queue is routed to the first available server. This is referred to as the serpentine model and is illustrated in the diagram given below.

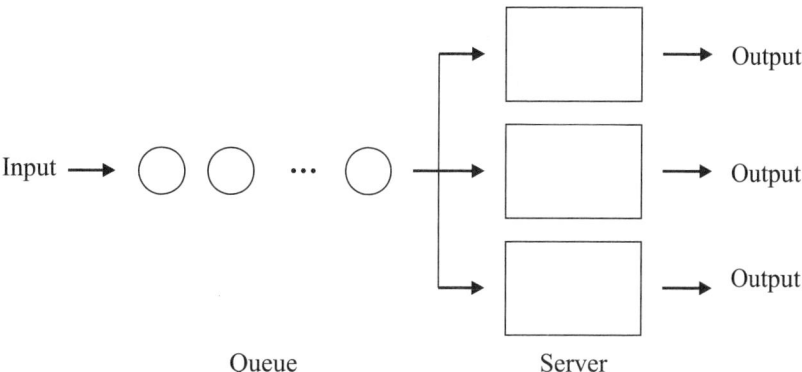

Managers study the behavior of different queuing systems in order to determine what level of service to provide. At a low level of service, the cost of providing service is low but the cost of customer dissatisfaction is high. At a high level of service, the cost of providing service is high but the cost of customer dissatisfaction is low. Careful analysis is required in order to determine the optimal level of service.

The cost of providing service is relatively easy to compute. However, the cost of customer dissatisfaction is more complicated. It is generally a function of various measures of the quality of service, including:

1. The probability that a customer or unit has to wait for service.
2. The waiting time.
3. The length of time in the system. (The length of time in the system includes the service time as well as the waiting time.)
4. The number of customers or units in the queue.
5. The number of customers or units in the system. (The number of customers or units in the system includes those being served as well as those in the queue.)

Formulas for these measures have been derived analytically for a number of different queuing systems. However, simulation is used to analyze the more complex queuing systems.

Please visit `http://www.bized.ac.uk/fme/5.htm` for additional information about queuing theory.

Compound Interest/Exponential Growth

Discrete Compounding

First we will define some variables.

P = Present value of money

F = Future value of money

r = Annual interest rate

t = Number of years of the investment

It is our goal to derive a formula for the future value of P dollars invested for t years at an interest rate r, compounded annually. After one year, the value of the investment is given by

F = Initial investment + interest earned for one year

$= P + Pr$

$= P(1+r)$.

Now we will reinvest the money for another year.

F = the value of the investment at the end of the first year + the interest earned for the year

$F = P(1+r) + P(1+r) \cdot r$ Factor out $P(1+r)$.

$= P(1+r)(1+r)$ Simplify the equation.

$= P(1+r)^2$

We will repeat this process again — we reinvest our money for another year (this will be the third year).

F = The value of the investment at the end of the second year + the interest earned for the year

$F = P(1+r)^2 + P(1+r)^2 \cdot r$ Factor out $P(1+r)^2$.

$= P(1+r)^2 (1+r)$ Simplify the equation.

$= P(1+r)^3$

Now we will summarize the results and see if there is a pattern that we can recognize.

The value of the investment after one year is $F = P(1+r)^1$.

The value of the investment after two years is $F = P(1+r)^2$.

The value of the investment after three years is $F = P(1+r)^3$.

Can you see a pattern? The exponent is equal to the number of years of the investment. So, if P dollars is invested for t years at an interest rate r and interest is paid once a year then

$$F = P(1+r)^t.$$

Example 1: Suppose that $500 is invested for 25 years at an interest rate of 6.0%, compounded annually. What is the future value of the investment?

SOLUTION

$$F = 500(1+0.06)^{25}$$
$$= 500(1.06)^{25}$$
$$= \$2{,}145.94$$

The initial investment earned $1,645.94 in 25 years. Instead of earning 6.0%, compounded annually, what if the interest was compounded semiannually, quarterly, monthly, or daily? Do you think that your investment would earn more money?

Now we will derive a new formula for the future value of money when interest is compounded n times a year. If the annual interest rate is r and interest is paid n times per year, then r/n will be the interest rate paid on the principal every time that interest is compounded. If $n = \dfrac{\text{number of compounding periods}}{\text{year}}$, then the total number of compounding periods = nt.

For example, if the principal is compounded semiannually (twice a year) for three years, then the total number of compounding periods = $nt = 2 \cdot 3 = 6$.

Recall that the future value of an investment of P dollars for t years at an interest rate r paid once a year is

$$F = P(1+r)^t.$$

As was previously stated, when interest is paid more frequently than once a year, the rate paid in each compounding period is r/n and the number of compounding periods is nt. Consequently, if P dollars are invested for t years at an interest rate of r, compounded n times a year, then the future value is given by

$$F = P\left(1+\frac{r}{n}\right)^{nt}.$$

In other words, the formula becomes

$$\text{Future value} = \text{Present value} \cdot \left[1+\frac{\text{interest rate}}{\text{\# of compounding periods}}\right]^{(\text{\# of compounding periods in } t \text{ years})}.$$

Example 2: Recall from **Example 1** that the future value of money was $2,145.94 for an investment of $500 for 25 years at an interest rate of 6.0%, compounded once a year. Let us compare this result with four different scenarios: semiannual compounding ($n = 2$), quarterly compounding ($n = 4$), monthly compounding ($n = 12$), and daily compounding ($n = 365$).

SOLUTION

a. Semiannual Compounding

$$F = 500\left(1+\frac{0.06}{2}\right)^{2 \cdot 25}$$
$$= 500(1.03)^{50}$$
$$= \$2{,}191.95$$

b. Quarterly Compounding

$$F = 500\left(1+\frac{0.06}{4}\right)^{4 \cdot 25}$$
$$= 500(1.015)^{100}$$
$$= \$2{,}216.02$$

c. Monthly Compounding

$$F = 500\left(1 + \frac{0.06}{12}\right)^{12 \cdot 25}$$
$$= 500(1.005)^{300}$$
$$= \$2,232.48$$

d. Daily Compounding

$$F = 500\left(1 + \frac{0.06}{365}\right)^{365 \cdot 25}$$
$$= 500(1.000164384)^{9125}$$
$$= \$2,240.57$$

In the previous examples we have been asked to find the future value of money given the present value of money, the interest rate, the number of compounding periods per year, and the number of years the money will be invested. There are, however, different applications that ask for variables other than the future value of money.

One situation that may arise is one in which the interest rate is the unknown variable. We will use the equation for the **future value** of money and solve for the interest rate. Pay close attention to the algebraic manipulations that are used to solve for the interest rate, r. The formula is difficult to memorize; however, if you understand the algebra you always can use the future value of money equation and the algebraic techniques to solve for the interest rate.

$$F = P\left(1 + \frac{r}{n}\right)^{nt}$$ Multiply both sides of the equation by $\frac{1}{P}$.

$$\frac{1}{P} \cdot F = P\left(1 + \frac{r}{n}\right)^{nt} \cdot \frac{1}{P}$$ Simplify the equation.

$$\frac{F}{P} = \left(1 + \frac{r}{n}\right)^{nt}$$ Take the nt root of both sides of the equation.

$$\left(\frac{F}{P}\right)^{1/(nt)} = \left[\left(1 + \frac{r}{n}\right)^{nt}\right]^{1/(nt)}$$ Use the property $\left(a^k\right)^{1/k} = a$.

$$\left(\frac{F}{P}\right)^{1/(nt)} = 1 + \frac{r}{n}$$ Subtract one from both sides of the equation.

$$\left(\frac{F}{P}\right)^{1/(nt)} - 1 = \frac{r}{n}$$ Multiply both sides of the equation by n.

$$n\left[\left(\frac{F}{P}\right)^{1/(nt)} - 1\right] = \frac{r}{n} \cdot n$$ Simplify the equation.

$$n\left[\left(\frac{F}{P}\right)^{1/(nt)} - 1\right] = r$$

The interest rate is given by

$$r = n\left[\left(\frac{F}{P}\right)^{1/(nt)} - 1\right].$$

We now have a formula for the interest rate given the future and present value of money, the number of compounding periods per year, and the length of time the money is invested.

Example 3: Adriana wants to invest $20,000. At what interest rate (to the nearest 0.01%), compounded semiannually, must she invest her money for five years in order for her to have $2,500 more than her initial investment?

SOLUTION

$$r = 2\left[\left(\frac{22,500}{20,000}\right)^{1/(2\cdot 5)} - 1\right]$$
$$= 2\left[(1.125)^{1/10} - 1\right]$$
$$= 0.023695$$
$$\cong 2.37\%$$

Adriana needs to find some type of investment that gives a 2.37% return on her money if her initial investment is to grow to $22,500 in five years when interest is compounded twice a year.

Logarithms

Another situation that you could encounter is one in which the unknown variable is time, t. Once again, the future value of money equation will be used to solve for time. Remember to pay attention to the algebraic manipulations. Before we begin, some logarithmic properties are needed.

Recall from algebra the properties of exponents. These properties are related to the logarithmic properties. Notice the relationships between the properties given in the table below.

Exponent	**Logarithm**
$a^m a^n = a^{m+n}$	$\log_a(pq) = \log_a p + \log_a q$
$\dfrac{a^m}{a^n} = a^{m-n}$	$\log_a\left(\dfrac{p}{q}\right) = \log_a p - \log_a q$
$(a^m)^n = a^{m \cdot n}$	$\log_a(p)^q = q \log_a p$

We also can use the relationship between the logarithm with base e and the natural logarithm.

$$\log_e x = \ln x$$

Now we can find the formula for t, the number of years of the investment, using some of the properties of logarithms. Once again, we will start with the **future value** of money equation.

$F = P\left(1 + \dfrac{r}{n}\right)^{nt}$ Multiply both sides of the equation by $\dfrac{1}{P}$.

$\dfrac{1}{P} \cdot F = P\left(1 + \dfrac{r}{n}\right)^{nt} \cdot \dfrac{1}{P}$ Simplify the equation.

$\dfrac{F}{P} = \left(1 + \dfrac{r}{n}\right)^{nt}$ Take the natural logarithm of both sides of the equation.

Compound Interest/Exponential Growth

$$\ln\left(\frac{F}{P}\right) = \ln\left(1+\frac{r}{n}\right)^{nt} \quad \text{Apply the logarithm property } \log_a(p)^q = q\log_a p.$$

$$\ln\left(\frac{F}{P}\right) = nt\ln\left(1+\frac{r}{n}\right) \quad \text{Divide both sides of the eqution by } n\ln\left(1+\frac{r}{n}\right).$$

$$\frac{\ln\left(\frac{F}{P}\right)}{n\ln\left(1+\frac{r}{n}\right)}$$

The time is given by

$$t = \frac{\ln\left(\frac{F}{P}\right)}{n\ln\left(1+\frac{r}{n}\right)}.$$

We now have a formula for the time, t, given the future and present value of money, the number of compounding periods per year, and the interest rate.

Example 4: Art wants to buy a car in five years. Since the stock market is too risky, he is going to invest his money at the local bank. The bank is offering a rate of 4%, compounded quarterly, on an investment of $5,000 or more in a certificate of deposit that will mature in five years. Assuming that Art will not have any money for his car other than the investment, will he have enough money when the CD matures to purchase a car that costs $5,500?

SOLUTION

$$t = \frac{\ln\left(\frac{5,500}{5,000}\right)}{4\cdot\ln\left(1+\frac{0.04}{4}\right)}$$

$$= \frac{\ln(1.1)}{4\cdot\ln(1.01)}$$

$$\cong 2.39$$

Art will definitely have enough money and more because it will take approximately 2.39 years for his initial investment of $5,000 to reach $5,500. The CD matures in five years so we can calculate the future value of the $5,000 investment, which will be the amount of money that Art will have to purchase a car.

$$F = P\left(1+\frac{r}{n}\right)^{nt}$$

$$= 5,000\left(1+\frac{0.04}{4}\right)^{4\cdot 5}$$

$$= 5,000(1.01)^{20}$$

$$= \$6,100.95$$

Continuous Compounding

Suppose that interest is compounded continuously. Do you think there would be a significant difference in the **future value** of money from the one that was compounded monthly?

Thus far we have only dealt with discrete compounding problems. Now we introduce continuous compounding.

We start with the formula for discrete compounding.

$$F = P\left(1 + \frac{r}{n}\right)^{nt} \quad \text{Substitute } n = kr.$$

$$F = P\left(1 + \frac{r}{kr}\right)^{krt} \quad \text{Manipulate the equation.}$$

$$F = P\left[\left(1 + \frac{1}{k}\right)^k\right]^{rt}$$

As n goes to infinity, so does kr. Since r is fixed, this implies that k goes to infinity as n goes to infinity.

Notice as $k \to \infty$ the graph of $F(k) = \left(1 + \frac{1}{k}\right)^k$ approaches e, where $e \approx 2.71828186...$.

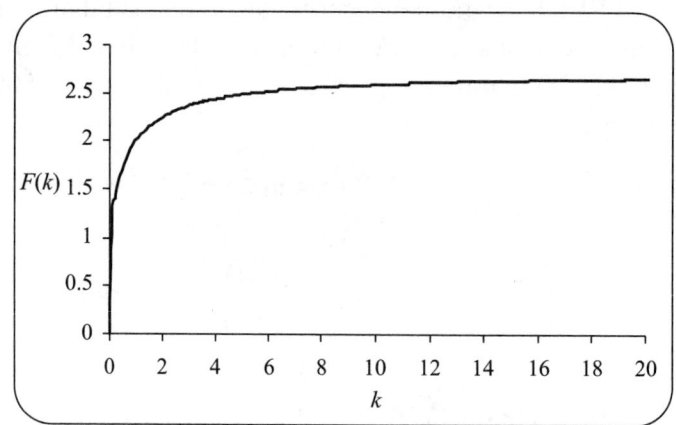

Therefore, as $k \to \infty$,

$$P\left[\left(1 + \frac{1}{k}\right)^k\right]^{rt} = P[e]^{rt}.$$

Now we have an equation for the future value of money when interest is compounded continuously.

$$F = Pe^{rt}$$

Example 5: Earlier in this section we compared four different scenarios; $500 was invested at 6.0% compounded annually, semiannually, quarterly, monthly, and daily for 25 years. A summary of the results is shown in the table below.

Compound Interest/Exponential Growth

Frequency of Compounding	Future value
Annual	$2,145.94
Semiannually	$2,191.95
Quarterly	$2,216.02
Monthly	$2,232.48
Daily	$2,240.57

Now we compute the future value of the investment when interest is compounded continuously.

SOLUTION

$$F = Pe^{rt}$$
$$= 500e^{0.06 \cdot 25}$$
$$= \$2,240.84$$

Because we have rounded to the nearest cent there appears to be a difference of $0.28 or 28 cents between the effects of continuous compounding and daily compounding. Why is this? It sure seems that you would be earning a lot more money if interest were compounded continuously rather than daily. The answer is quite simple.

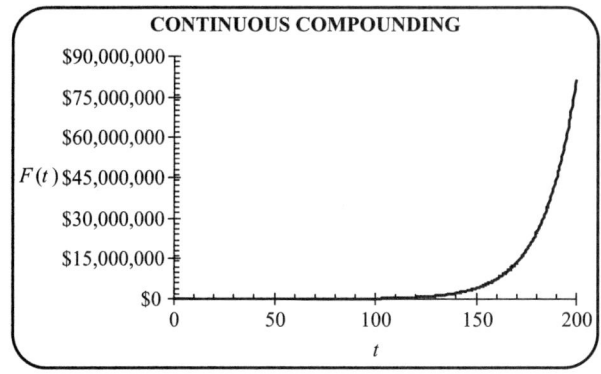
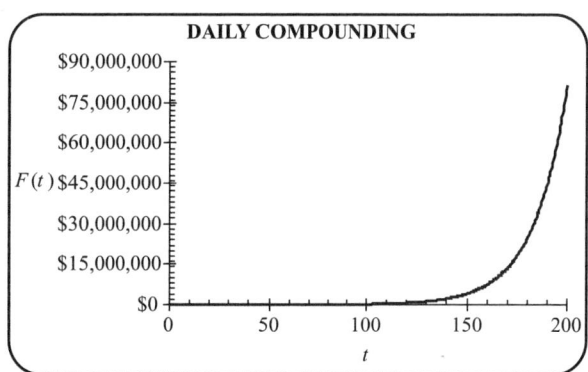

$$F(t) = Pe^{rt} \qquad\qquad F(t) = P\left(1 + \frac{r}{n}\right)^{nt}$$

Earlier we discovered that as $n \to \infty$

$$F = P\left(1 + \frac{r}{n}\right)^{nt} = Pe^{rt}.$$

This means that, for a large number of compounding periods, discrete compounding will closely approximate continuous compounding. Therefore, the future value of the investment will never be greater than Pe^{rt}. If you look at the graph on page 94 you will see that for large values of k, $F(k)$ approaches e. What about small values of k? You will notice that for small values of k, the value of the function approaches e very quickly. How does this idea relate to the formulas for discrete and continuous compounding? Large values of k are analogous to large values of n. Hence, for large values of n, discrete compounding is approximately equal to continuous compounding. Conversely, small values of k are analogous to small values of n; therefore, for small values of n, the future values for discrete compounding quickly approach the future values for continuous compounding. This is why there is not much difference between daily compounding and continuous compounding as illustrated in the two graphs above. We can also analyze the two graphs by examining the table below.

t	Daily Compounding	Continuous Compounding	Difference
1	$530.92	$530.92	$0.00
2	$563.74	$563.75	$0.01
3	$598.60	$598.61	$0.01
4	$635.61	$635.62	$0.01
5	$674.91	$674.93	$0.02
6	$716.64	$716.66	$0.02
7	$760.95	$760.98	$0.03
8	$808.01	$808.04	$0.03
9	$857.97	$858.00	$0.04
10	$911.01	$911.06	$0.04
15	$1,229.71	$1,229.80	$0.09
20	$1,659.89	$1,660.06	$0.16
25	$2,240.57	$2,240.84	$0.28
30	$3,024.38	$3,024.82	$0.45
35	$4,082.38	$4,083.08	$0.70
40	$5,510.50	$5,511.59	$1.09
45	$7,438.22	$7,439.87	$1.65
50	$10,040.29	$10,042.77	$2.48
55	$13,552.64	$13,556.32	$3.68
60	$18,293.70	$18,299.12	$5.41
65	$24,693.31	$24,701.22	$7.92
70	$33,331.66	$33,343.17	$11.51
75	$44,991.92	$45,008.57	$16.64
80	$60,731.25	$60,755.21	$23.96
85	$81,976.59	$82,010.95	$34.37
90	$110,654.09	$110,703.21	$49.12
95	$149,363.72	$149,433.70	$69.98
100	$201,614.96	$201,714.40	$99.44
150	$4,048,546.37	$4,051,541.96	$2,995.59
200	$81,297,181.43	$81,377,395.71	$80,214.28
250	$1,632,495,000.79	$1,634,508,686.24	$2,013,685.45
300	$32,781,455,405.99	$32,829,984,568.67	$48,529,162.68

Notice that the difference between daily compounding and continuous compounding (where $P = \$500$ and $r = 6.0\%$) for $t \leq 65$ is less than $10.00. For $65 < t \leq 100$, the difference is less than $100. For $t > 100$, the difference starts to grow rapidly. At 300 years the difference between compounding daily and continuously is $48,529,162.68. In college algebra you learned that this is characteristic of exponential growth.

As seen with discrete compounding, the situation may arise where the unknown variables are the interest rate, r, or the time, t. Using properties of logarithms we can solve for both of these variables. Remember to pay attention to the steps involved.

$F = Pe^{rt}$ Divide both sides of the equation by P.

$\dfrac{F}{P} = e^{rt}$ Take the natural logarithm of both sides of the equation.

$\ln\left(\dfrac{F}{P}\right) = \ln(e^{rt})$ Apply the logarithm property $\ln e^p = p$.

$\ln\left(\dfrac{F}{P}\right) = rt$ Divide each side of the equation by r or t to obtain the following equations.

$$t = \dfrac{\ln\left(\dfrac{F}{P}\right)}{r} \quad \text{and} \quad r = \dfrac{\ln\left(\dfrac{F}{P}\right)}{t}$$

Example 6: If $8,000 is compounded continuously at 2%, how long will it take for the money to grow to $9,000? Round your answer to the nearest day.

SOLUTION

$$t = \dfrac{\ln\left(\dfrac{9,000}{8,000}\right)}{0.02}$$

$\cong 5.88915$ years

$= 5 \text{ years } + 0.88915 \cdot \dfrac{365 \text{ days}}{1 \text{ year}}$

$\cong 5$ years 325 days

Example 7: At what annual interest rate will $12,000 grow to $14,000 in three years if interest is compounded continuously?

SOLUTION

$$r = \dfrac{\ln\left(\dfrac{14,000}{12,000}\right)}{3}$$

$= 0.051383$

$\cong 5.14\%$

Effective Yield

Let us revisit **Example 2**, the example in which we investigated different frequencies of compounding. Recall that $P = \$500$, $r = 6.0\%$, $n = 1, 2, 4, 12$, and 365, and $t = 25$ years. The results are shown below.

Compounded	Future value
Annually	$2,145.94
Semiannually	$2,191.95
Quarterly	$2,216.02
Monthly	$2,232.48
Daily	$2,240.57

Note that the future values when interest is compounded semiannually, quarterly, monthly, and daily ($n = 2, 4, 12$, and 365) are greater than the future value when interest is compounded annually ($n = 1$). What if we wanted to find an annual rate compounded **once a year** for t years that would produce the same future value if the principal was invested at an annual rate, r, compounded n times a year for t years? This new annual rate is called the **effective yield**, y.

The formula for the **effective yield** is derived in the following calculations.

Future Value (compounded n times for one year)
    = *Future Value* (compounded once a year for one year)

$$P\left(1+\frac{r}{n}\right)^n = P(1+y) \quad \text{Divide both sides of the equation by } P.$$

$$\left(1+\frac{r}{n}\right)^n = 1+y \quad \text{Subtract 1 from both sides of the equation.}$$

$$\left(1+\frac{r}{n}\right)^n - 1 = y$$

The effective yield is given by

$$y = \left(1+\frac{r}{n}\right)^n - 1.$$

Now we compute the effective yield for the investment that was compounded semiannually in **Example 2**.

$$y = \left(1+\frac{0.06}{2}\right)^2 - 1$$
$$= 0.0609$$
$$= 6.09\%$$

The annual rate that is actually earned on the investment is 6.09%, not 6.0%. If the interest rate had been 6.09%, compounded annually, then the future value of the investment would be the same as the future value when interest rate is 6.0%, compounded semiannually. The computations are

Compound Interest/Exponential Growth

shown below.

$$F = 500(1+0.0609)^{25}$$
$$= \$2{,}191.95$$

Note that this future value is equivalent to the result of $F = 500\left(1+\dfrac{0.06}{2}\right)^{2 \cdot 25}$.

The table given below shows the effective yields when interest is compounded annually, semiannually, quarterly, and daily.

Compounded	Interest rate	Effective yield
Annually	6.0%	6.0000%
Semiannually	6.0%	6.0900%
Quarterly	6.0%	6.1364%
Monthly	6.0%	6.1678%
Daily	6.0%	6.1831%

You can see that the effective yield is greater than the quoted rate. Once again this is because compounding is not taken into consideration.

Example 8: A bank is offering an annual interest rate of 3%, compounded monthly, on three-year certificates of deposit. What is the effective yield on this investment?

SOLUTION

$$y = \left(1+\frac{0.03}{12}\right)^{12} - 1$$
$$\cong 0.0304$$
$$= 3.04\%$$

The formula for the effective yield when interest is compounded continuously is derived the same way as discrete compounding.

Future Value (compounded continuously for 1 year)
$\quad = $ *Future Value* (compounded once a year for 1 year)

$Pe^r = P(1+y) \quad$ Divide both sides of the equation by P.
$e^r = 1+y \quad$ Subtract 1 from both sides of the equation.
$e^r - 1 = y$

The effective yield is given by

$$y = e^r - 1.$$

Example 9: The effective yield can be used to compare two investments. One bank is offering an annual rate of 5%, compounded continuously, on two-year certificates of deposit. A second bank is offering an annual rate of 5%, compounded weekly, on one-year certificates of deposit. Which bank is offering the better rate?

SOLUTION

$$\text{Bank 1:} \quad y = e^r - 1 \qquad\qquad \text{Bank 2:} \quad y = \left(1 + \frac{r}{n}\right)^n - 1$$

$$= e^{0.05} - 1 \qquad\qquad\qquad\qquad = \left(1 + \frac{0.05}{52}\right)^{52} - 1$$

$$\cong 0.0513 \qquad\qquad\qquad\qquad\qquad \cong 0.0512$$

$$= 5.13\% \qquad\qquad\qquad\qquad\qquad\quad = 5.12\%$$

The effective yield on the two-year certificate of deposit is 0.01% higher than the effective yield on the one-year certificate of deposit. However, in order to earn the slightly higher effective yield, the money would need to be kept in the bank for an additional year. If the money will be needed after the end of the first year but before the end of the second year, then the one-year certificate of deposit with an effective yield of 5.12% would be the more desirable investment.

Another question can be asked related to the effective yield. What is the annual rate r that would produce a given effective yield? This is found by solving for the annual rate in the effective yield equation. Pay close attention to the algebraic steps involved.

Discrete compounding:

$y = \left(1 + \frac{r}{n}\right)^n - 1$ Add 1 to both sides of the equation.

$y + 1 = \left(1 + \frac{r}{n}\right)^n$ Take the nth root of both sides of the equation.

$(y+1)^{1/n} = 1 + \frac{r}{n}$ Subtract 1 from both sides of the equation.

$(y+1)^{1/n} - 1 = \frac{r}{n}$ Multiply both sides of the equation by n.

$n\left[(y+1)^{1/n} - 1\right] = r$

The annual rate r, compounded n times per year, that would produce an effective rate y is

$$r = n\left[(y+1)^{1/n} - 1\right].$$

Continuous compounding:

$y = e^r - 1$ Add one to both sides of the equation.

$y + 1 = e^r$ Take the natural logarithm of both sides of the equation.

$\ln(y+1) = \ln(e^r)$ Use the logarithm property $\ln e^p = p$.

$\ln(y+1) = r$

Compound Interest/Exponential Growth

The annual rate r, compounded continuously, that would produce an effective rate y is
$$r = \ln(y+1).$$

Example 10: Customers of the Arizona Bank earn an effective yield of 4% when interest is compounded quarterly. What is the annual interest rate, r?

SOLUTION

$$r = n\left((y+1)^{1/n} - 1\right)$$
$$= 4\left((0.04+1)^{1/4} - 1\right)$$
$$\cong 0.0394$$
$$= 3.94\%$$

The annual interest rate is 3.94%, compounded quarterly.

Example 11: A competitor, Premiere Bank, is offering the same effective yield as the Arizona Bank (in **Example 10**) with continuous compounding. Which bank is offering the better annual rate, r?

SOLUTION

$$r = \ln(y+1)$$
$$= \ln(0.04+1)$$
$$\cong 0.0392$$
$$= 3.92\%$$

The Arizona Bank is offering the higher annual rate of 3.94%. Does the answer make sense? Yes, because if interest is compounded continuously, then a lower quoted rate is required in order to produce the same future value as that earned when interest is compounded quarterly. This is illustrated in the next example with the use of the formulas for the future value of money when interest is compounded semiannually and when interest is compounded continuously.

Example 12: At what rate does $10,000 need to be invested in order to have $10,500 in two years?
  Case 1: Interest is compounded semiannually.

SOLUTION

$$F = P\left(1 + \frac{r}{n}\right)^{nt}$$

$$10,500 = 10,000\left(1 + \frac{r}{2}\right)^{2 \cdot 2}$$

$$\frac{10,500}{10,000} = \left(1 + \frac{r}{2}\right)^4$$

$$\left(\frac{10,500}{10,000}\right)^{1/4} = \left[\left(1+\frac{r}{2}\right)^4\right]^{1/4}$$

$$\left(\frac{10,500}{10,000}\right)^{1/4} = 1+\frac{r}{2}$$

$$\left(\frac{10,500}{10,000}\right)^{1/4} - 1 = \frac{r}{2}$$

$$2\left[\left(\frac{10,500}{10,000}\right)^{1/4} - 1\right] = r$$

$$0.0245 \cong r$$

$$2.45\% = r$$

Case 2: Interest is compounded continuously.

As discussed before, the interest rate should be smaller when compounding twice a year as compared to continuous compounding.

SOLUTION

$$F = Pe^{rt}$$

$$10,500 = 10,000e^{r \cdot 2}$$

$$\ln\left(\frac{10,500}{10,000}\right) = \ln\left(e^{2r}\right)$$

$$\ln\left(\frac{10,500}{10,000}\right) = 2r$$

$$\frac{\ln\left(\frac{10,500}{10,000}\right)}{2} = r$$

$$0.0244 = r$$

$$2.44\% = r$$

Value of Money

What if there was a situation in which we needed to find the present value of money? First we need to find a formula for the **present value** of money. We will use the equation for the **future value** of money for discrete compounding.

$$F = P\left(1+\frac{r}{n}\right)^{nt}$$

Multiply both sides of the equation by $\dfrac{1}{\left(1+\dfrac{r}{n}\right)^{nt}}$.

Compound Interest/Exponential Growth

$$\frac{1}{\left(1+\frac{r}{n}\right)^{nt}} \cdot F = P\left(1+\frac{r}{n}\right) \cdot \frac{1}{\left(1+\frac{r}{n}\right)^{nt}} \quad \text{Simplify the equation.}$$

$$\frac{1}{\left(1+\frac{r}{n}\right)^{nt}} \cdot F = P \quad \text{Use the exponent property } \frac{1}{a^n} = a^{-n}.$$

$$\left(1+\frac{r}{n}\right)^{-nt} \cdot F = P$$

The present value of money is given by

$$P = F\left(1+\frac{r}{n}\right)^{-nt}.$$

We can also find the **present value** of money for continuous compounding.

$$F = Pe^{rt} \quad \text{Divide both sides of the eqution by } e^{rt}.$$

$$\frac{F}{e^{rt}} = P \quad \text{Use the exponent property } \frac{1}{a^n} = a^{-n}.$$

$$Fe^{-rt} = P$$

The present value of money for continuous compounding is given by

$$P = Fe^{-rt}.$$

Example 13: A newborn child, named Brian, had a trust fund set up by his grandparents for his future education. The trust fund is guaranteed to produce a return of 5% compounded quarterly until the child reaches his 18th birthday. On Brian's 18th birthday he discovers that he has $24,459.21 to use for college, thanks to his grandparent's gift. How much money did Brain's grandparents put into his trust fund when he was born?

SOLUTION

$$F = \$24{,}459.20, \quad r = 0.05, \quad t = 18, \quad n = 4.$$

$$P = F\left(1+\frac{r}{n}\right)^{-nt}$$

$$= 24{,}459.21\left(1+\frac{0.05}{4}\right)^{-4 \cdot 18}$$

$$= 24{,}459.21(1.0125)^{-72}$$

$$= \$10{,}000.$$

Example 14: Answer the question in **Example 13** but assume interest is compounded continuously.

SOLUTION

$$P = Fe^{-rt}$$
$$= 24{,}459.21e^{-0.05(18)}$$
$$= \$9{,}944.37$$

There are some business applications where the ratio, R, of future to present value of money is used for continuous compounding. We start with the equation for the future value of money for continuous compounding, $F = Pe^{rt}$, and divide both sides by P.

$$R = \frac{F}{P} \qquad \text{Substitute } Pe^{rt} \text{ for } F.$$
$$= \frac{Pe^{rt}}{P}$$
$$= e^{rt}$$

The ratio of the future value to the present value when interest is compounded continuously is

$$R = e^{rt}.$$

Example 15: Compute the yearly ratio of future to present value of money that corresponds to an annual rate of 4.5% compounded continuously.

SOLUTION

$$R = e^{rt}$$
$$= e^{0.045 \cdot 1}$$
$$\cong 1.046$$

Logarithms — Another look

There are two crucial facts that are especially important to keep in mind when working with logarithms:

1) The $\log_a x$ is a power; specifically, it is the power to which a needs to be raised to obtain x. Mathematically, $y = \log_a x$ means the same as $a^y = x$.
2) The $\log_a x$ is the inverse of a^x. Mathematically, $\log_a a^x = x$ for all x and $a^{\log_a x} = x$ for all $x > 0$.

To facilitate discussion, please keep in mind the following terminology for the expression $y = \log_a x$. The "base" of $y = \log_a x$ is a, the "argument" or "input" is x, and the logarithm itself, the power, is y.

Let's see these facts in mathematical action.

Example 16:
   a. Given $\log_2 8 = y$, solve for y.
   b. Given $\log_2 16 = y$, solve for y.

SOLUTION

a. $\log_2 8 = 3$ because 3 is the power to which 2 needs to be raised to obtain 8. Mathematically, $2^3 = 8$.

b. $\log_2 16 = 4$ because 4 is the power to which 2 needs to be raised to obtain 16. Mathematically, $2^4 = 16$.

Similarly, $\log_7 49 = 2$ because $7^2 = 49$ and $\log_7 \frac{1}{49} = -2$ because $7^{-2} = \frac{1}{49}$. Recall that negative exponents reciprocate ("one over") their arguments. That is, $a^{-r} = \frac{1}{a^r}$. Also recall that fractional powers are equivalent to roots; that is, $a^{1/3} = \sqrt[3]{a}$. In general, this is $a^{1/n} = \sqrt[n]{a}$.

Example 17:

a. What is $\log_2 \sqrt[5]{2}$?

b. What is $\log_3 \frac{1}{\sqrt[7]{3}}$?

SOLUTION

a. $\log_2 \sqrt[5]{2} = \frac{1}{5}$ because $2^{1/5} = \sqrt[5]{2}$.

b. $\log_3 \frac{1}{\sqrt[7]{3}} = -\frac{1}{7}$ because $3^{-1/7} = \frac{1}{\sqrt[7]{3}}$.

There are couple more things to keep in mind when you are working with logarithms. First, when working with logarithms that have base 10, the base is omitted from the notation; so $\log_{10} x$ is written as $\log x$ and is called a "common" logarithm. Second, when working with logarithms that have base e, the base is omitted from the notation and log is replaced with ln, so we have $\log_e x = \ln x$.

Remember: logarithms are powers!

Now, how about the second crucial fact? Logarithms are inverses for exponentials with the same base. Remember back in elementary school when you first learned addition? What was the next major topic? I'll wager it was subtraction. How about later in elementary school when you first learned how to multiply? What was the next topic that you learned in your mathematics class? Let me consult my crystal ball; yes, it looks like division! What do these pairs of operations have in common? Addition and subtraction are inverse operations. Likewise, multiplication and division are also inverses of each other.

If you have $10 and I give you $2, then turn around and take away $2, what is the final result of my actions? Other than perhaps annoy you, nothing. $(10 + 2) - 2 = 10$.

The same is true for a^x and $\log_a x$. These functions are inverses. They "undo" each other.

Example 18: Solve each logarithmic expression for the unknown variable.

a. $\log_{20} 20^8 = y$

b. $\log_{20} 20^{-12} = y$

c. $\log_{20} 20^{6.321} = y$
d. $\log 10^{-3} = y$
e. $\log 1{,}000{,}000 = y$
f. $\log \sqrt[7]{10} = y$
g. $\ln e^6 = y$

SOLUTION

a. $\log_{20} 20^8 = 8$
b. $\log_{20} 20^{-12} = -12$
c. $\log_{20} 20^{6.321} = 6.321$
d. $\log 10^{-3} = -3$
e. $\log 1{,}000{,}000 = \log 10^6 = 6$
f. $\log \sqrt[7]{10} = \log 10^{1/7} = \dfrac{1}{7}$
g. $\ln e^6 = \log_e e^6 = 6$

Note that since the domain of a^x is all real numbers, the inverse property, as demonstrated above, works for any input x; so the $\log_a a^x = x$ for any real number x. We can also use this inverse process in the other direction. However, as you may or may not remember, the logarithm function has a domain restriction. Namely, you can input only positive numbers into the logarithm function. Thus, $5^{\log_5 18} = 18$ as you would expect from the fact that exponential functions and logarithm functions are inverses of each other; but $5^{\log_5 -18} \neq -18$ because -18 is not in the domain of $\log_5 x$. Similarly, $e^{\ln 3.7} = 3.7$, but $e^{\ln 0} \neq 0$ because 0 is not in the domain of $\ln x$.

Before we move onto solving equations involving logarithmic or exponential expressions, let's first review a basic mathematical concept that applies to all levels of mathematics from basic arithmetic to vector calculus and beyond. Have you ever heard a mathematics teacher use the acronym, PEMDAS? Or "Please excuse my dear Aunt Sally"? Regardless of what mathematics you are doing, you must always be careful to follow proper mathematical protocol, called **order of operations**.

P Parenthesis first
E Exponential expressions (and logarithms)
M Multiplication*
D Division*
A Addition*
S Subtraction*

* PEMDAS is a bit misleading, as multiplication and division actually share the same level on the order of operation, as do addition and subtraction. To decide which of two competing operations to perform first, if the levels of operation are the same, the left-to-right rule is invoked. Note that if an addition is competing with a multiplication, however, it does not matter if the addition is to

the left of the multiplication or not; the multiplication is performed first, as it enjoys higher level on the order of operation protocol. Also, don't forget that when working with nested parentheses, or parentheses within parentheses, to perform the operations in the innermost set first and then work your way out.

Example 19: Simplify $7 + 8^2 - 100 \div 10 \cdot 5 + 6(3 - 5)$.

SOLUTION

$7 + 8^2 - 100 \div 10 \cdot 5 + 6(-2)$	P	Parenthesis first
$7 + 64 - 100 \div 10 \cdot 5 + 6(-2)$	E	Exponential expressions
$7 + 64 - 10 \cdot 5 + 6(-2)$	M/D	Multiplication/Division (L-to-R)
$7 + 64 - 50 + (-12)$	M/D	Multiplication/Division (L-to-R)
$71 - 50 + (-12)$	A/S	Addition/Subtraction (L-to-R)
$21 + (-12)$	A/S	Addition/Subtraction (L-to-R)
9	A/S	Addition/Subtraction (L-to-R)

Example 20: Simplify $10(3 - 2(5 + 4(6 - 2)))$.

SOLUTION

$10(3 - 2(5 + 4 \cdot 4))$	P	Innermost parenthesis first
$10(3 - 2(5 + 16))$	M	Multiplication (2nd level parenthesis)
$10(3 - 2 \cdot 21)$	A	Addition (2nd level parenthesis)
$10(3 - 42)$	M	Multiplication (3rd level parenthesis)
$10(-39)$	S	Subtraction (3rd level parenthesis)
-390	M	Multiplication

Now that we have dusted off our order of operation memory cells, we can utilize proper order of operations, together with the basic properties of exponential and logarithmic expressions, to solve equations involving such expressions. Remember that when solving an equation for an unknown, use the necessary inverse operations, but in reverse order.

Example 21: Solve for x: $y = 1 - e^{x/\alpha}$.

SOLUTION

First let's break down what is happening to x on the right-hand side of this exponential equation.

1st x is divided by α (note that $e^{x/\alpha}$ is actually $e^{\wedge}(x \div \alpha)$, so the division is performed before the exponentiation due to the implied parentheses),

2nd e is raised to the x/α power,

3rd $e^{x/\alpha}$ is negated; and finally

4th 1 is added to $-e^{x/\alpha}$.

To solve for x, we undo each of the above operations *in reverse order*. Also, remember that when solving equations, whatever is done to one side of the equation must be done to both sides.

Undo step 4 by subtracting 1 from both sides of the equation:
$$y - 1 = 1 - e^{x/\alpha} - 1$$
$$y - 1 = -e^{x/\alpha}$$

Undo step 3 by multiplying both sides by (–1):
$$-(y - 1) = -(-e^{x/\alpha})$$
$$1 - y = e^{x/\alpha}$$

Undo step 2 by taking the natural logarithm of both sides:
$$\ln(1 - y) = \ln e^{x/\alpha}$$
$$\ln(1 - y) = x/\alpha$$

Undo step 1 by multiplying both sides by α:
$$\alpha \ln(1 - y) = \alpha(x/\alpha)$$
$$\alpha \ln(1 - y) = x$$

Example 22: Solve $3 \cdot 2^{3x-5} - 10 = 38$ for x.

SOLUTION

$$3 \cdot 2^{3x-5} - 10 + 10 = 38 + 10$$
$$3 \cdot 2^{3x-5} = 48$$
$$\frac{3 \cdot 2^{3x-5}}{3} = \frac{48}{3}$$
$$2^{3x-5} = 16$$
$$\log_2 2^{3x-5} = \log_2 16$$
$$3x - 5 = 4$$
$$3x - 5 + 5 = 4 + 5$$
$$3x = 9$$
$$\frac{3x}{3} = \frac{9}{3}$$
$$x = 3$$

Verify answer: $3 \cdot 2^{3 \cdot 3 - 5} - 10 = 3 \cdot 2^4 - 10 = 3 \cdot 16 - 10 = 48 - 10 = 38$

Example 23: Solve $\log(x + 2) + \log(x + 4) = 3$ for x.

SOLUTION

$$\log_2(x + 2) + \log_2(x + 4) = 3$$
$$\log_2(x + 2)(x + 4) = 3$$
$$(x + 2)(x + 4) = 2^3$$
$$x^2 + 6x + 8 = 8$$
$$x^2 + 6x = 0$$
$$x(x + 6) = 0$$
$$x = 0 \text{ or } x = -6$$

Note that −6 is not in the domain of either $\log_2(x+2)$ or $\log_2(x+4)$; $(-6+2)$ and $(-6+4)$ both result in negative values, neither which are in the domain of $\log_2 x$. Thus, $x = 0$ is the only legitimate solution.

Application

An interesting application of logarithms that we are all familiar with is the Richter scale for measuring the magnitude of earthquakes. The Richter scale is a base-10 (common log) logarithmic scale. The input into the logarithmic function is the output from a seismometer, which measures the amplitude of the seismic waves.

An earthquake of magnitude 6 on the Richter scale can cause damage, but primarily to poorly constructed buildings within a 10 kilometer radius of the epicenter. A magnitude 9 earthquake, as occurred on December 26, 2004 in the Pacific Ocean, can cause hideous and widespread destruction and massive death tolls. The difference between a relatively harmless quake and a violently destructive one is only 3 points on the Richter scale. This initially seems surprising, until we remember that the Richter scale is a logarithmic scale with base 10. Thus an increase of one on the Richter scale is actually a power of 10 increase. An increase of 3 points on the Richter scale, therefore, corresponds to an earthquake $10^3 = 1000$ times more powerful. Imagine the difference between dropping a 1-pound book on your foot, perhaps causing you to hop around, cursing a bit, and a 1000-pound piece of equipment falling on your foot.

Example 24: How many times more powerful is a magnitude 8 "great quake" than a magnitude 4 quake, which is often felt, but rarely does damage?

SOLUTION

A magnitude 8, on the Richter scale, earthquake is $10^4 = 10,000$ times more powerful than a magnitude 4 earthquake.

While logarithmic modeling fits some real-world situations perfectly, it is the exponential model which works in a wide variety of situations. It is quite common, for instance, for quantities to vary over time by percent change as opposed to varying by a constant amount. We would not expect the population of New York City to change by the same amount each year as the population of Berea, Kentucky, but the percent change might be similar. Quantities that vary by a constant **percent** change per unit time can be modeled with the exponential model: $Q(t) = Q_0 e^{kt}$, where Q_0 is the quantity present at "time zero" and k is the continuous growth rate. Do you recall what types of functions are used to model quantities that vary by a constant **amount** per unit time? These are the linear functions, $y = mx + b$.

Example 25: According to an article entitled, "Arctic ice cap may melt by Summer 2070," by Michael Hennigan, Jan. 2, 2005 (www.finfacts.com/irelandbusinessnews), the Arctic ice is only half as thick as it was just 30 years ago. Additionally, the distribution of the ice has been reduced by 10% during the same period. This would represent a 55% loss of volume over 30 years. If we assume the percent decay rate is constant, approximately when will there remain only 10% of the present volume of ice in the Arctic ice cap?

SOLUTION

Using the exponential decay model, $Q = Q_0 e^{-kt}$, we first need to determine the continuous decay rate k; then we find the t for which $Q = 0.10 Q_0$. Note that $0.10 Q_0$ represents 10% of the present volume of ice. As this equation involves an exponential function, we will need the inverse of an exponential, a logarithmic function, to solve for t.

As there has been a 55% loss over 30 years, only 45% of the ice from 30 years ago is still present, or $0.45 Q_0 = Q_0 e^{-k \cdot 30}$. Solving this equation for k, we obtain:

$$0.45 = e^{-30k}$$
$$\ln 0.45 = \ln e^{-30k}$$
$$\ln 0.45 = -30k$$
$$\frac{\ln 0.45}{-30} = k$$
$$0.0266 \cong k.$$

Substituting $k \cong 0.0266$ into our model, we have:

$$0.10 Q_0 = Q_0 e^{-0.0266t}$$
$$0.10 = e^{-0.0266t}$$
$$\ln(0.10) = \ln e^{-0.0266t}$$
$$\ln(0.10) = -0.0266t$$
$$\frac{\ln(0.10)}{-0.0266} = t$$
$$86.6 \cong t.$$

If the exponential rate of decay remains constant, then by approximately the summer of 2091 only 10% of the Arctic ice cap will remain. Note that the article voices the concern that the ice cap will vanish by the summer of 2070. If in fact the rate of decay is increasing, this may be a reasonable prediction.

Economic forecasting and future market analyses usually require predicting future demographic distribution. The U.S. Census Bureau (www.census.gov) gathers information about present demographics through direct counting, but higher level mathematics is needed to form predictions about what the future will hold.

Example 26: According to the U.S. Census Bureau, the U.S. Hispanic or Latino origin population grew from 35,306,000 in 2000 to 39,899,000 in 2003. During the same time period, the White, non-Hispanic, population grew from 195,576,000 to 197,326,000. If these growth patterns continue, when will the U.S. Hispanic population and the U.S. White population have the same size?

SOLUTION

While initially one may look at this large population difference and assume it will be centuries before the Hispanic population catches up to the White population in the United State, one should not so quickly dismiss the power of exponential growth.

To solve this problem, we need to use the exponential population model, $P = P_0 e^{kt}$, to determine the population continuous growth rates of both populations. Note that we are using $t = 0$ to represent the year 2000, so $t = 3$ will correspond to the year 2003.

Hispanic population:

$$39899000 = 35306000 e^{3k}$$

$$\frac{39899000}{35306000} = e^{3k}$$

$$\ln \frac{39899}{35306} = \ln e^{3k}$$

$$0.1223 \cong 3k$$

$$\frac{0.1223}{3} \cong k$$

$$0.04077 \cong k$$

White population (using a similar mathematical process):

$$197,326,000 = 195,576,000 e^{3k}$$

$$0.002969 \cong k$$

Now we input our continuous growth rates into their respective models and set them equal to each other (resulting in future populations of equal size) and solve for t.

$$35306000 e^{0.04077t} = 195576000 e^{0.002969t}$$

$$e^{(0.04077 - 0.002969)t} = \frac{195576000}{35306000}$$

$$e^{0.03781t} \cong 5.539$$

$$\ln e^{0.03781t} \cong \ln 5.539$$

$$0.03781t \cong \ln 5.539$$

$$t \cong \ln 5.539 / 0.03781$$

$$t \cong 45.28$$

By the year 2045, if the respective growth rates continue on their present trends, we should expect the Hispanic and White populations in the United States to be approximately of equal size.

Exercises

1. Suppose that $2,000 is needed at the end of four years and that money can be deposited into an account that pays interest at a rate of 5%.
   a. How much money must be deposited if interest is compounded monthly?
   b. What is the effective yield on the account if interest is compounded monthly?
   c. How much money must be deposited if interest is compounded continuously?
   d. What is the effective yield on the account if interest is compounded continuously?

SOLUTION

a. $$F = P\left(1+\frac{r}{n}\right)^{nt}$$

$$2000 = P\left(1+\frac{0.05}{12}\right)^{12 \cdot 4}$$

$$P = 2000\left(1+\frac{0.05}{12}\right)^{-12 \cdot 4}$$

$$= \$1,638.14$$

b. $$y = \left(1+\frac{r}{n}\right)^n - 1$$

$$= \left(1+\frac{0.05}{12}\right)^{12} - 1$$

$$\cong 0.0512$$

$$= 5.12\%$$

c. $$F = Pe^{rt}$$

$$2000 = Pe^{0.05 \cdot 4}$$

$$P = 2000e^{-0.05 \cdot 4}$$

$$= \$1,637.46$$

d. $$y = e^r - 1$$

$$= e^{0.05} - 1$$

$$\cong 5.13\%$$

2. How long will it take $10,000 to grow to $15,000 if interest is paid at an annual rate of 2.5%, compounded continuously?

SOLUTION

$$F = Pe^{rt}$$

$$15,000 = 10,000e^{0.025t}$$

$$\frac{15,000}{10,000} = e^{0.025t}$$

$$\ln\left(\frac{15,000}{10,000}\right) = \ln\left(e^{0.025t}\right)$$

$$\ln\left(\frac{15,000}{10,000}\right) = 0.025t$$

$$t = \frac{\ln\left(\frac{15,000}{10,000}\right)}{0.025} \cong 16.22 \text{ years}$$

Compound Interest/Exponential Growth

3. What interest rate, compounded semiannually, is needed for a $5,000 investment to grow to $6,500 in 5 years?

SOLUTION

$$r = n\left(\left(\frac{F}{P}\right)^{1/(nt)} - 1\right)$$

$$= 2\left(\left(\frac{6,500}{5,000}\right)^{1/(2\cdot 5)} - 1\right)$$

$$= 2\left((1.3)^{1/10} - 1\right)$$

$$\cong 5.32\%$$

4. What rate, r, will produce an effective yield of 4% when compounded continuously?

SOLUTION

$$r = \ln(1+y)$$
$$= \ln(1+0.04)$$
$$= \ln(1.04)$$
$$\cong 3.92\%$$

5. Find the present value of a $650,000 payment 20 years from now if interest is earned at a rate of 4.25% compounded quarterly.

SOLUTION

$$P = 650,000\left(1 + \frac{0.0425}{4}\right)^{-4\cdot 20}$$

$$= 650,000(1.010625)^{-80}$$

$$= \$279,068.22$$

6. If interest is compounded monthly, what interest rate is needed for a $1,000 investment to double in ten years?

SOLUTION

$$F = P\left(1 + \frac{r}{n}\right)^{nt}$$

$$2000 = 1000\left(1 + \frac{r}{12}\right)^{12\cdot 10}$$

$$2 = \left(1 + \frac{r}{12}\right)^{120}$$

$$(2)^{1/120} = \left[\left(1+\frac{r}{12}\right)^{120}\right]^{1/120}$$

$$(2)^{1/120} = 1+\frac{r}{12}$$

$$12\left[(2)^{1/120} - 1\right] = r$$

$$0.0695 \cong r$$

$$6.95\% = r$$

7. At what rate does $500 need to be invested in order to have $550 in three years if interest is compounded semiannually? Round your answer to three decimal places.

SOLUTION

$r = 3.202\%$

8. Repeat Exercise 7 but use continuous compounding.

SOLUTION

$r = 3.177\%$

9. A bank offers three different Certificates of Deposit (CD): 3.5% for two years, 4.0% for three years, and 4.5% for five years that are compounded quarterly. You would like to have $3,000 when the CD matures. Because you will need the money to purchase a car in three years, you will choose the CD that matures in three years. How much money would you have to invest? Round your answer to the nearest dollar.

SOLUTION

$P = \$2,662$ for the three-year CD

10. How long will it take $8,000 to grow to $10,000 if interest is paid at an annual rate of 4.75% compounded semiannually? Express your answer to the nearest month.

SOLUTION

$t \cong 4.75$ years $= 4$ years 9 months

11. How much will $3,500 be worth in 10 years if the annual interest rate is 3.25% compounded continuously? Round your answer to the nearest cent.

SOLUTION

$F = \$4,844.11$

12. Explain the difference between the annual interest rate, r, and the effective yield, y.

Compound Interest/Exponential Growth

SOLUTION

The annual interest rate, r, compounded n times per year, is the rate at which your principal, P, is invested for t years to obtain a future value of F.

The effective yield, y, is the rate, **compounded annually**, at which your principal, P, is invested for t years to obtain the same future value, F, when P is invested for t years at an interest rate of r, compounded n times a year.

13. Compute the effective yield in Exercise 11 and round your answer to five decimal places. Show that the yield will give the same future value, after 10 years, as Exercise 11 if interest is compounded annually for 10 years.

SOLUTION

$y = 3.30339\%$

$F = 3500 \cdot \left(1 + \dfrac{0.0330339}{1}\right)^{1 \cdot 10}$

$= \$4{,}844.11$

14. Compute the yearly ratio of future to present value of money that corresponds to an annual rate of 2.25% compounded continuously.

SOLUTION

$R \cong 1.02$

15. True or False? The effective yield, y, will always be greater than the annual interest rate, r, for discrete or continuous compounding. Explain your answer.

SOLUTION

True

16. Find the following logarithms (try not to use a calculator: remember, logarithms are powers!).

a. $\log_5 25$ d. $\log_3 \dfrac{1}{81}$ g. $\log 1$

b. $\log 10{,}000$ e. $\log_2 \sqrt{2}$ h. $\ln 1$

c. $\log_3 81$ f. $\log 0.001$ i. $\ln(-1)$

SOLUTION

Logarithms are powers.

a. 2   d. −4   g. 0

b. 4   e. $\dfrac{1}{2}$ h. 0

c. 4   f. −3   i. $\ln(-1)$ is undefined.

17. Simplify the following expressions using the fact that logarithms and exponential functions are inverses of each other, provided the base is the same for both expressions.
   a. $\log_4 4^7$
   b. $\log_2 2^{-3}$
   c. $\log 10^{52}$
   d. $\ln e^{-1.734}$
   e. $e^{\ln 72}$
   f. $10^{\log 9.735}$
   g. $3^{\log_3 (-4)}$

SOLUTION

Logarithms and exponential functions, of the same base, are inverses of each other and thus undo each other under the operation of composition of functions.
   a. 7
   b. −3
   c. 52
   d. −1.734
   e. 72
   f. 9.735
   g. $3^{\log_3 (-4)}$ is undefined.

18. Solve $2 \cdot e^{3x+1} - 9 = 31$ for x. (Give your answer accurate to 3 decimal places.)

SOLUTION

$$x = \frac{-1 + \ln 20}{3} \cong 0.665$$

19. Solve $F = Pe^{rt}$ for t.

SOLUTION

$$t = \frac{1}{r} \cdot \ln \frac{F}{P}$$

20. Solve $5\ln(2x - 1) = 10$ for x. (Give your answer accurate to 3 decimal places.)

SOLUTION

$$x = \frac{e^2 + 1}{2} \cong 4.195$$

21. Solve $5 \cdot 3^{2x-9} + 10 = 145$.

SOLUTION

$$x = 6$$

22. Solve $\log(x - 1) - \log(x + 1) = -1$ for x. (Hint: First combine the logarithms on the left-hand side of the equation using basic logarithm properties.)

SOLUTION

$$x = \frac{11}{9}$$

Compound Interest/Exponential Growth

23. Approximately how many times more powerful is an earthquake of magnitude 9.5 on the Richter scale than an earthquake of magnitude 4.5? (Refer to the application used for **Example 24**.)

SOLUTION

$10^5 = 100{,}000$ times

24. Using the exponential decay model in **Example 25**, what would the continuous decay rate need to be for the Artic ice cap to melt to 1% of its present (2005) volume by 2070?

SOLUTION

7.08%

25. Using the information from **Example 26**, what would the U.S. Hispanic population and the White population be in 2045?

SOLUTION

The Hispanic population will be approximately 221 million, and the White population will be 223.5 million.

26. Given that the U.S. Asian population in 2000 was approximately 10,589,000, what would the Asian population continuous growth rate need to be to have the same population size as the U.S. White population in 2045? Use 223.5 million for the White population in 2045.

SOLUTION

Approximately 6.78%

27. According to a recent article in the business section of an Arizona newspaper, cereal prices are rising exponentially with a continuous growth rate of approximately 8%. If this trend continues, how much would a $3 box of cereal cost in 20 years?

SOLUTION

Approximately $14.86

28. Using the information from Exercise 27, approximately when will the $3 box of cereal cost $15?

SOLUTION

Approximately 2025

Histograms

Histograms organize data into groups by counting the number of observations in each group. Using formal terminology, groups are referred to as **bins** and the number of observations that are in a bin is referred to as the **frequency**. For example, suppose that we would like to sort the data given below into bins.

4.000000	0.282449	2.729575
0.498367	0.688040	0.653401
3.585772	0.510575	2.450636
1.344646	4.443800	1.533403
1.413617	4.894711	2.876827

You first need to decide how many bins to use and how large each bin should be. Since the above observations are all between 0 and 5, five bins will be used. When working with a large amount of data, use between six and fifteen bins and choose a convenient bin width.

Minimum	0.282449
Maximum	4.894711
Range	4.612262

$$Bin\ Width = \frac{Range}{\#\ of\ bins}$$
$$= \frac{Maximum - Minimum}{\#\ of\ bins}$$
$$= \frac{4.894711 - 0.282449}{5}$$
$$= 0.922452$$
$$\cong 1$$

Since this data set is small we can organize the data into bins manually. Notice the interval that corresponds to each of the bins.

Bin 1 (−∞, 1]	Bin 2 (1, 2]	Bin 3 (2, 3]	Bin 4 (3, 4]	Bin 5 (4, 5]
0.498367	1.344646	2.729575	4.000000	4.443800
0.282449	1.413617	2.450636	3.585772	4.894711
0.688040	1.533403	2.876827		
0.510575				
0.653401				

Now we can count the number of data points that are in each bin. The results are called the frequencies.

	Bin 1 (−∞, 1]	Bin 2 (1, 2]	Bin 3 (2, 3]	Bin 4 (3, 4]	Bin 5 (4, 5]
Frequencies →	5	3	3	2	2

The process of organizing this data into bins and computing the frequencies was simple since this data set is small. However, we need to use the *Histogram* function in *Excel* when working with a large amount of data.

Example 1: Use the *Histogram* function in *Excel* to sort the data in the table below into nine bins, and plot the results.

1.43	9.01	2.22	3.35	3.48	6.92	4.62	3.30	0.47	1.31
0.35	7.45	4.14	13.23	3.49	4.52	0.60	0.86	1.79	5.74
7.38	0.17	1.87	6.42	5.28	1.96	1.80	0.20	0.97	5.96
0.75	2.40	1.51	0.22	7.22	8.73	8.81	0.48	0.64	1.95
0.30	2.04	2.85	4.38	1.85	2.14	0.93	3.12	0.95	0.03
3.66	3.05	3.14	5.39	4.28	6.71	5.17	4.32	4.59	13.07
0.16	1.73	0.15	2.75	8.51	1.46	12.84	0.72	1.24	5.05
1.14	2.54	2.52	1.92	0.51	4.44	6.09	1.35	17.85	2.29
0.47	5.99	3.76	0.47	3.10	2.13	6.90	5.57	1.35	5.24
2.11	0.34	5.05	2.48	5.99	4.57	6.43	6.01	0.81	3.45

SOLUTION

First we must use *Excel* to calculate the minimum, maximum, range, and bin width.

Minimum	0.03
Maximum	17.85
Range	17.82
Bin Width	1.98

Now we introduce a new term—the **adjusted bin width**. The adjusted bin width is a convenient value for the bin width. For this example, the bin width is 1.98. A more convenient value would be 1 or 2. If we use an adjusted bin width of 1, then 18 bins would be needed in order to cover the interval [0.03, 17.85]. On the other hand, only 9 bins would be needed if we use an adjusted bin width of 2. Therefore, we must use an adjusted bin width of 2.

After we enter the appropriate bin limits into our *Excel* worksheet, we are ready to use the *Histogram* function to sort the data into the bins. Pull down the "Tools" menu and select "Data Analysis" and then "Histogram."

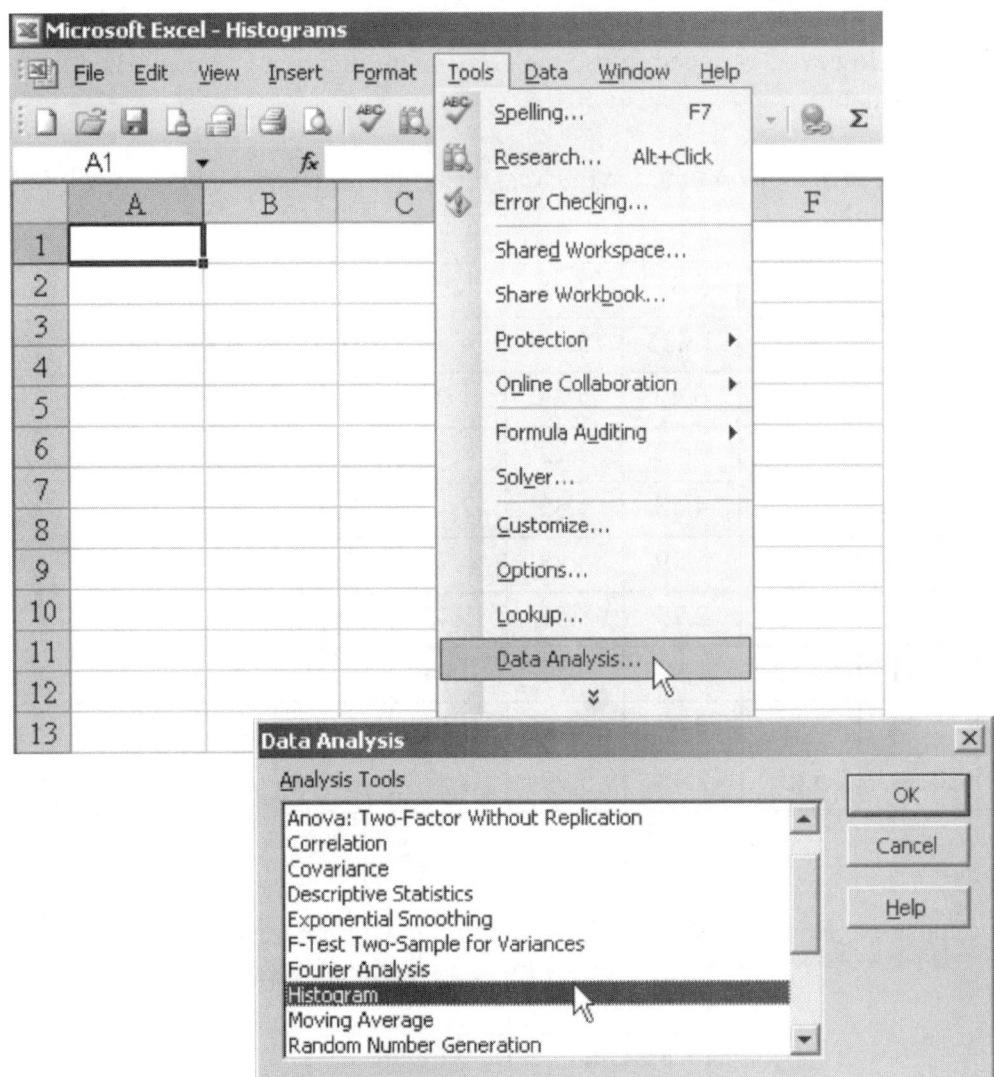

Then enter the appropriate information into the dialog box.

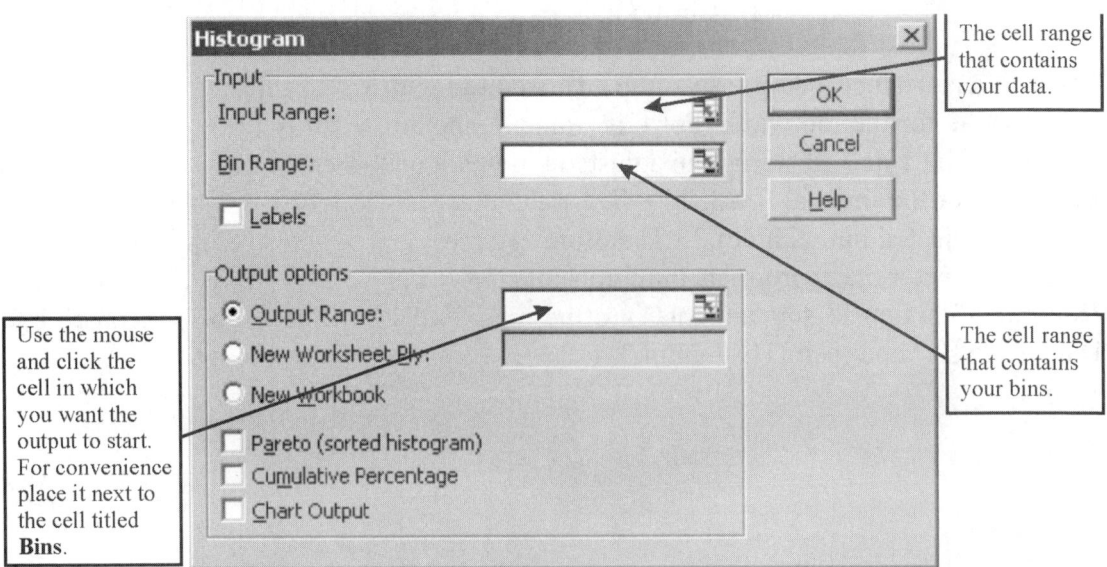

Click OK and you should see something similar to the table below. The first two columns were entered manually, and the last two columns were generated by the *Histogram* function. You can clear the cells that contain "More" and "0." Note that frequency counts the number of observations that are greater than the previous bin limit and less than or equal to the current bin limit. For example, there are 39 observations in the interval (0, 2] and there are 23 observations in the interval (2, 4].

Bin #	Bins	*Bin*	*Frequency*
1	2	2	39
2	4	4	23
3	6	6	20
4	8	8	10
5	10	10	4
6	12	12	0
7	14	14	3
8	16	16	0
9	18	18	1
		More	0

Try to reproduce the above table.

Now we can use the *Chart Wizard* in *Excel* to graph the results. Pull down the "Insert" menu and select "Chart."

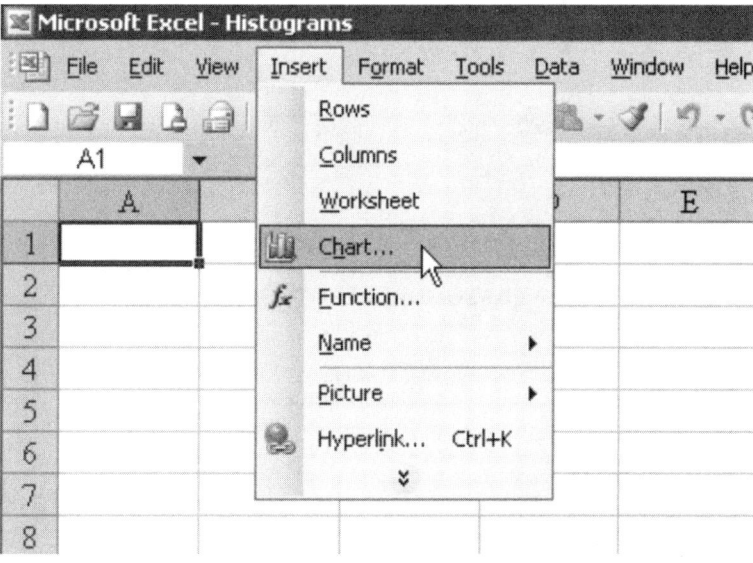

Then follow the steps shown below.

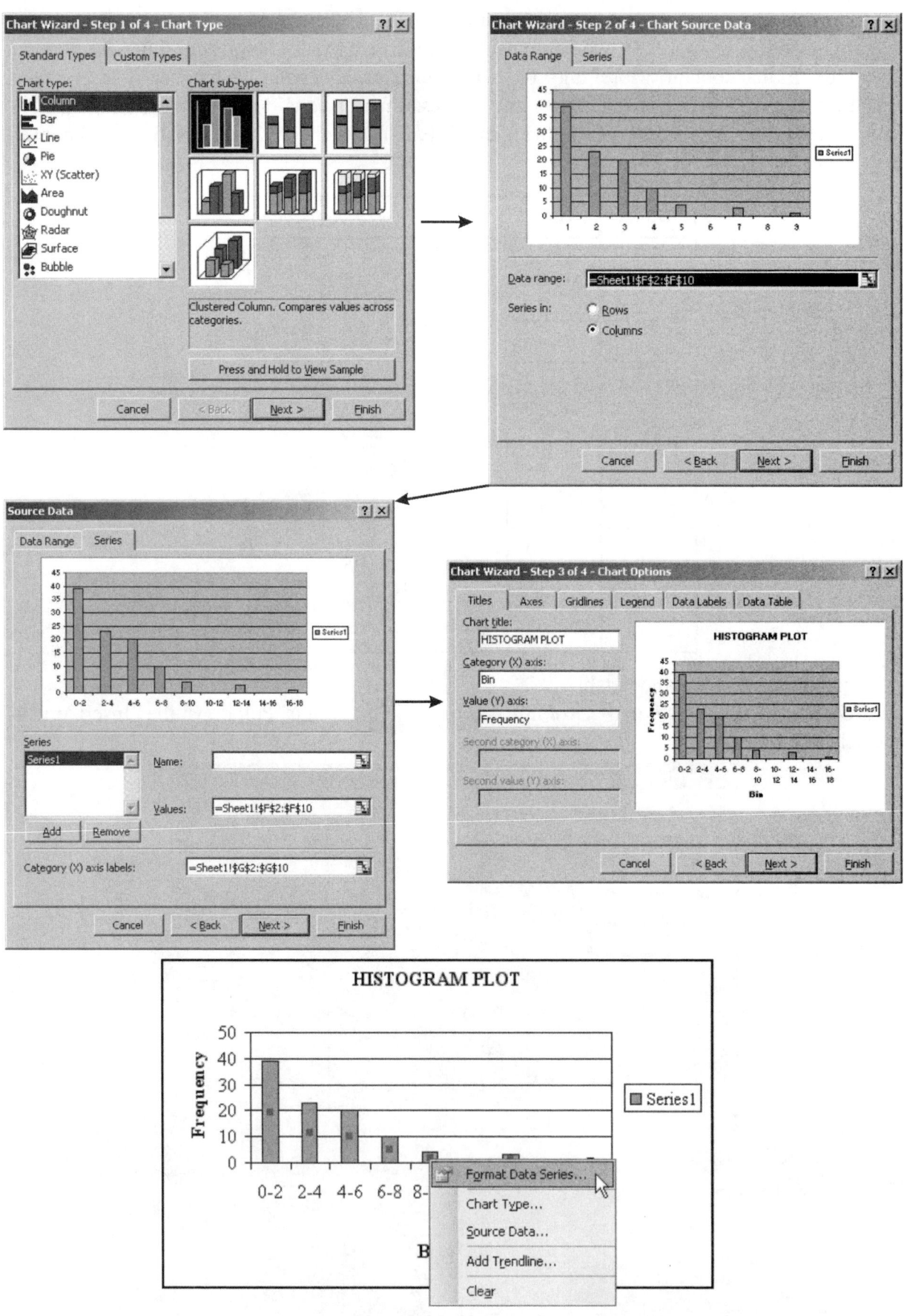

Histograms

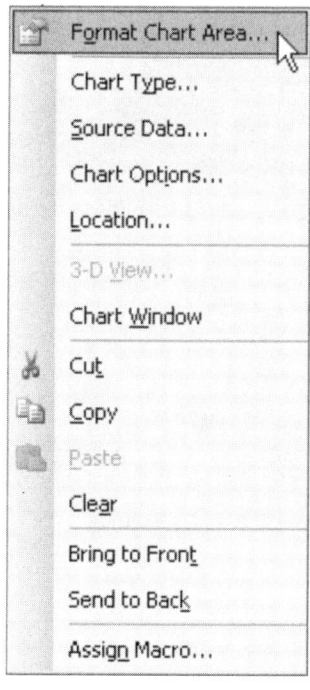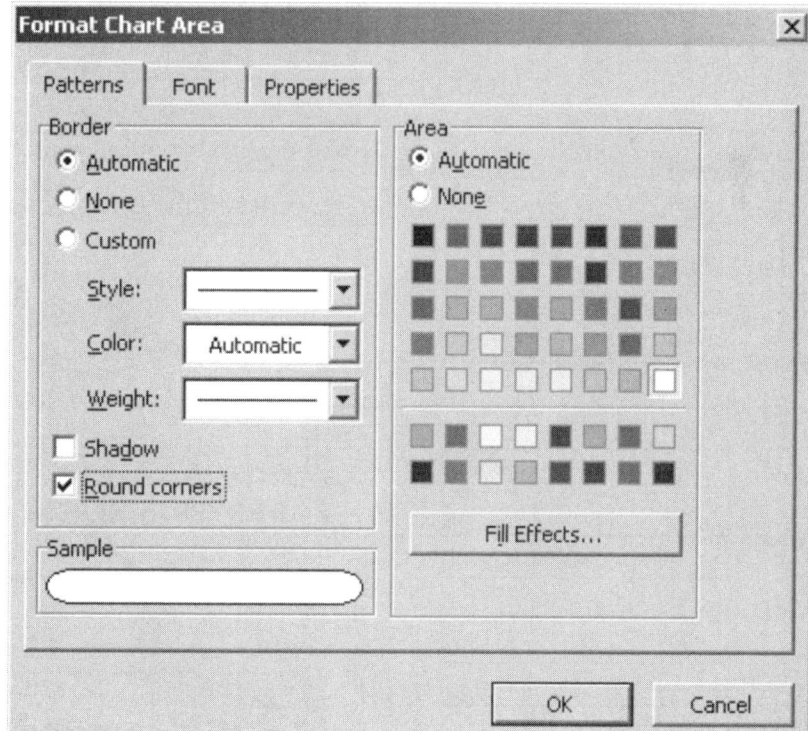

The final result is shown below.

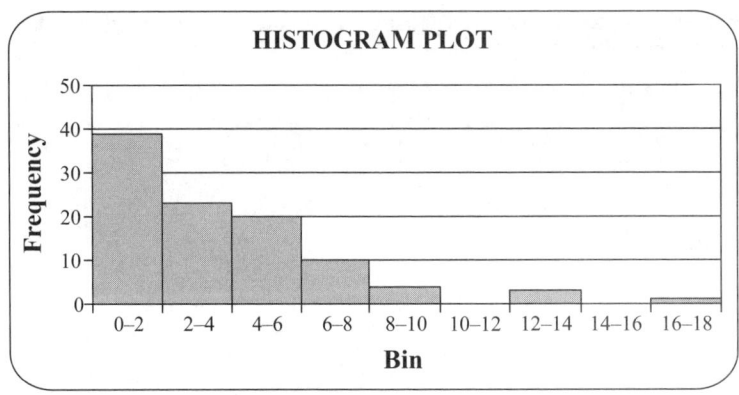

Example 2: The table below gives the estimated gross incomes (in millions of dollars) of the 40 highest paid entertainers in 1995 and 1996 combined.

130	31	43	44	44
51	59	50	77	33
25	46	90	40	30
32	40	56	33	50
63	75	35	74	42
31	44	28	59	36
63	44	28	48	171
33	28	36	150	42

Data obtained from `fortune.com`.

a. Find the minimum, maximum, and average estimated gross income of the entertainers.
b. Use the **Histogram** function in *Excel* to sort the data into eight groups and plot the results.

SOLUTION

a.

Minimum	Maximum	Average
25	171	53.4

b.

Bin	Frequency
40	16
60	15
80	5
100	1
120	0
140	1
160	1
180	1
Sum	40

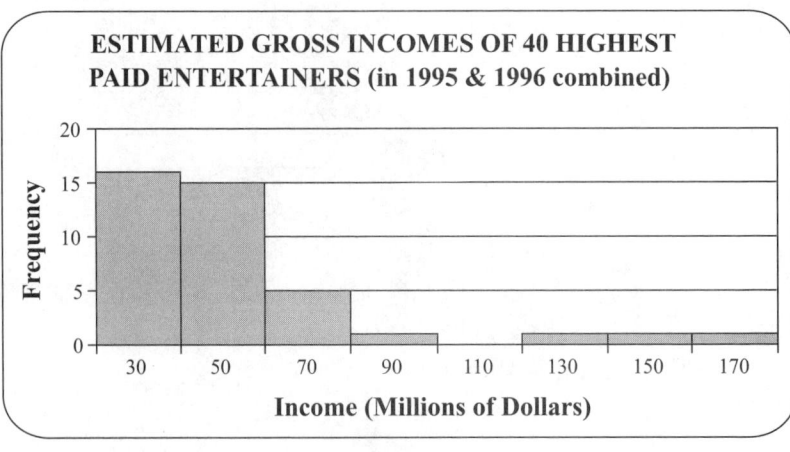

Thus far we have used the histogram function in *Excel* to sort the data into bins and to count the number of observations in each bin. In addition to the frequency, the relative frequency is of interest. **Relative frequency** is the ratio of the number of observations in a bin to the total number of observations. The formula for relative frequency is given below.

$$\text{Relative frequency} = \frac{\text{Frequency}}{\text{Sum of the Frequencies}}$$

Relative frequency can be used to compare the sizes of the bins in terms of percentages.

Example 3: In Example 2 we plotted the frequencies of the estimated gross incomes of the highest paid entertainers in 1995 and 1996 combined.
  a. Plot the relative frequencies of the estimated gross incomes.
  b. For what percentage of entertainers is the estimated gross income between 60 and 80 million dollars?
  c. What interval of estimated gross incomes has the largest relative frequency?

SOLUTION

a.

Bin	Frequency	Relative Frequency
40	16	40.0%
60	15	37.0%
80	5	12.5%
100	1	2.5%
120	0	0.0%
140	1	2.5%
160	1	2.5%
180	1	2.5%
Sum	40	100.0%

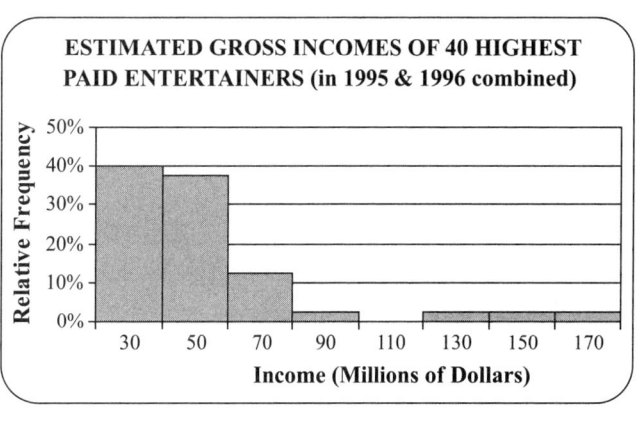

  b. The estimated gross income is between 60 and 80 million dollars for 12.5% of the entertainers.
  c. The interval from 20 to 40 million dollars has the largest relative frequency of 40%.

Example 4: Information about 37 individuals or families in Asia, included in the list of billionaires published by *Fortune* magazine in 1992, is given below. Wealth and age are reported in the table.

Wealth	14.0	4.9	4.0	4.0	3.9	3.9	3.8	3.8	3.6	3.4	3.3	3.0	2.8	2.5	2.4	2.3	2.2	2.0	1.8
Age	64	62	68	49	64	83	41	78	80	54	69	57	68	49	76	54	45	73	62
Wealth	1.8	1.8	1.8	1.8	1.7	1.7	1.6	1.6	1.6	1.4	1.4	1.3	1.2	1.2	1.2	1.2	1.2	1.0	
Age	68	60	53	67	53	67	62	69	78	52	73	59	69	59	68	69	64	69	

a. Find the minimum, maximum, and average age of the billionaires.
b. Use the *Histogram* function in *Excel* to sort the ages of the billionaires into nine groups (Hint: The bins are 45, 50, 55, 60, 65, 70, 75, 80, and 85.)
c. Compute the relative frequencies of the groups created in Part b.
d. Plot the relative frequencies of the ages of the billionaires.
e. What interval of ages has the largest relative frequency?
f. What interval of ages has the smallest relative frequency?

SOLUTION

a.

Minimum	Maximum	Average
41	83	63.6

b and c.

Bin	Frequency	Relative Frequency
45	2	5.41%
50	2	5.41%
55	5	13.51%
60	4	10.81%
65	6	16.22%
70	11	29.73%
75	2	5.41%
80	4	10.81%
85	1	2.70%

d.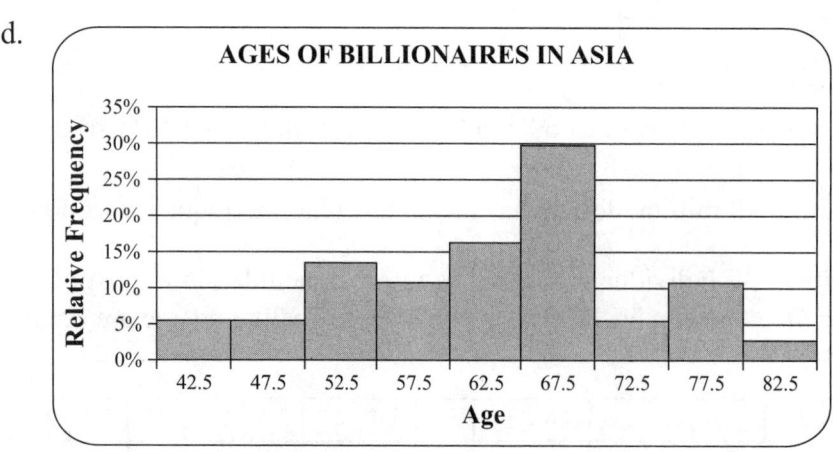

e. The interval from 65 to 70 years of age has the largest relative frequency of 0.2973.
f. The interval from 80 to 85 years of age has the smallest relative frequency of 0.0270.

Histograms

Exercises

1. Count the number of observations in the table below that are in each of the given bins.

39.8	25.8	32.1	27.9	3.6	15.0	52.1	35.2	47.9	18.6
6.1	56.7	40.9	55.6	41.2	3.9	24.6	1.8	20.1	60.0

Bin	Frequency
10	
20	
30	
40	
50	
60	

SOLUTION

Bin	Frequency
10	4
20	2
30	4
40	3
50	3
60	4

2. Sort the data in the table below into 10 bins and plot the frequencies.
   a. What bin has the largest frequency?
   b. What bin has the smallest frequency?

3.7208	1.8430	2.2117	1.3698	2.1929	0.0638	2.2344	0.4595	2.8111	3.4626
4.5305	1.3393	1.1362	2.7909	2.0425	2.0875	4.7131	2.5752	2.2098	4.8320
1.3205	4.4908	2.4694	2.3716	4.7316	3.2078	3.3448	2.0353	3.7640	2.3632
1.1542	0.6810	2.9035	0.8161	4.8689	2.2469	0.2809	1.8435	4.1539	1.7544
1.2830	0.4295	3.1202	2.1749	3.5804	1.3097	1.8230	3.0627	1.8516	2.9202
0.0334	2.3627	2.9875	4.9142	0.3180	0.2977	0.0529	4.3973	4.9561	1.7414
0.7923	3.5408	3.7252	0.9854	3.9541	3.7272	2.0891	0.3040	4.2221	3.8870
1.8149	1.2243	4.0292	3.1782	3.5549	4.9075	3.0386	0.6711	3.4213	0.5481
3.5708	2.1891	3.0874	3.1974	2.7612	1.9642	0.4447	4.9152	4.7247	4.2644
0.3485	0.6597	3.5144	2.6466	2.4821	2.8202	0.1497	1.9935	0.8707	0.7962

SOLUTION

Minimum	Maximum	Range	Bin Width	Adjusted Bin Width
0.0334	4.9561	4.9227	0.4923	0.5

Bin	Frequency
0.5	12
1.0	9
1.5	8
2.0	9
2.5	16
3.0	9
3.5	10
4.0	11
4.5	6
5.0	10
More	0

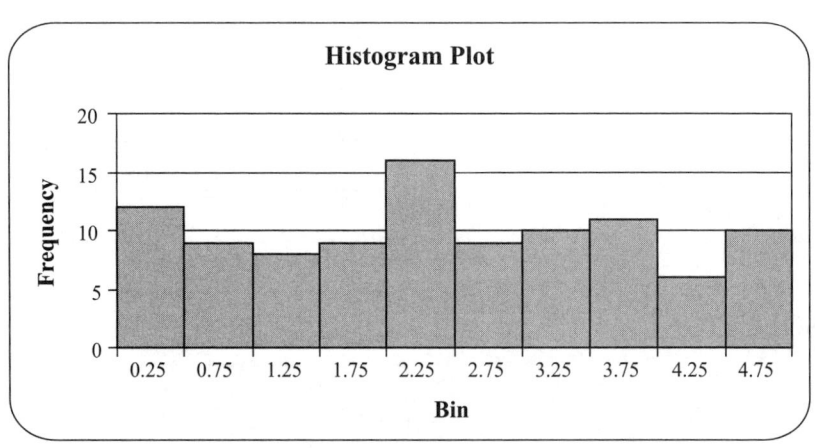

a. (1.5, 2.0]   b. (3.0, 3.5]

3. The table below gives the total amount of oil recovered (in thousands of barrels) from 64 wells in the Devonian Richmond Dolomite area of the Michigan basin.

21.71	56.4	43.4	36.6	79.5	12	82.2	12.1
53.2	49.4	69.5	64.9	26.9	28.3	35.1	20.1
46.4	44.9	156.5	14.8	18.5	204.9	47.6	30.5
42.7	34.6	34.6	17.6	14.7	44.5	54.2	7.1
50.4	92.2	37.9	29.1	32.9	10.3	63.1	10.1
97.7	37	12.9	61.4	196	37.7	69.8	18
103.1	58.8	2.5	38.6	24.9	33.7	57.4	3
51.9	21.3	31.4	32.5	118.2	81.1	65.6	2

J. Marcus Jobe and Hutch Jobe, "A statistical approach for additional infill development," *Energy Exploration and Exploitation*, 18 (2000), pp. 89–103.

a. Find the minimum, maximum, and average amount of oil recovered from the wells.
b. Use the *Histogram* function in *Excel* to sort the data into 14 groups, and compute the relative frequencies of the groups.
c. Plot the results from Part b.
d. What interval of amounts of oil has the largest relative frequency?

SOLUTION

a.

Minimum	Maximum	Average
2.0	204.9	48.2

b.

Bin	Frequency	Relative Frequency
15	11	17.1875%
30	10	15.6250%
45	17	26.5625%
60	10	15.6250%
75	6	9.3750%
90	3	4.6875%
105	3	4.6875%
120	1	1.5625%
135	0	0.0000%
150	0	0.0000%
165	1	1.5625%
180	0	0.0000%
195	0	0.0000%
210	2	3.1250%

c.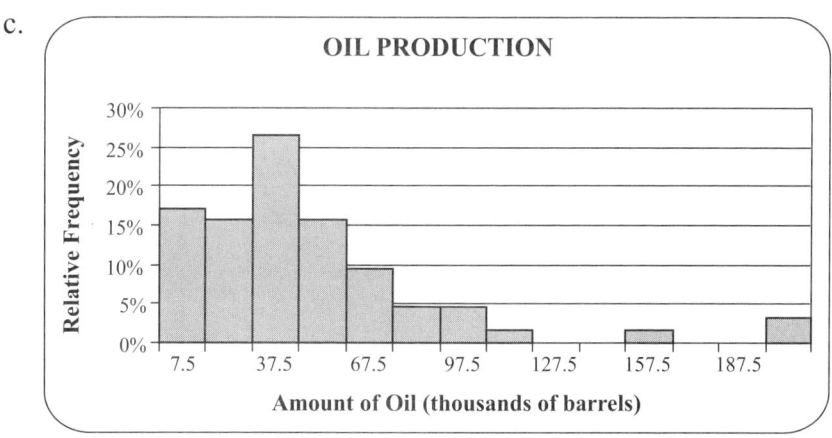

d. 30 to 45 thousand barrels of oil.

4. The table below gives the total assets (in billions of dollars) held by the world's 50 largest banking companies in 2001.

1281.4	1051.5	924.1	854.7	815.1	805.4	753.8	734.8	694.6	693.6
638.5	628.6	621.8	608.8	601.0	536.8	531.7	519.1	476.5	456.9
454.5	444.5	441.1	432.5	428.5	372.2	344.3	330.5	324.2	313.1
312.8	312.1	307.6	287.4	271.8	269.0	242.5	231.2	225.4	215.3
211.5	203.6	203.3	194.3	183.5	181.1	180.6	180.4	178.5	177.1

The World Almanac and Book of Facts. New York: World Almanac Books, 2003.

a. Find the minimum, maximum, and average total assets held by the banking companies.

b. Use the *Histogram* function in *Excel* to sort the data into 12 groups, and compute the relative frequencies of the groups.

c. Plot the results from Part b.

d. What interval of total assets has the largest relative frequency?

SOLUTION

a.

Minimum	Maximum	Average
177.1	1281.4	453.7

b.

Bin	Frequency	Relative Frequency
200	7	14%
300	10	20%
400	8	16%
500	7	14%
600	3	6%
700	7	14%
800	2	4%
900	3	6%
1000	1	2%
1100	1	2%
1200	0	0%
1300	1	2%

c.

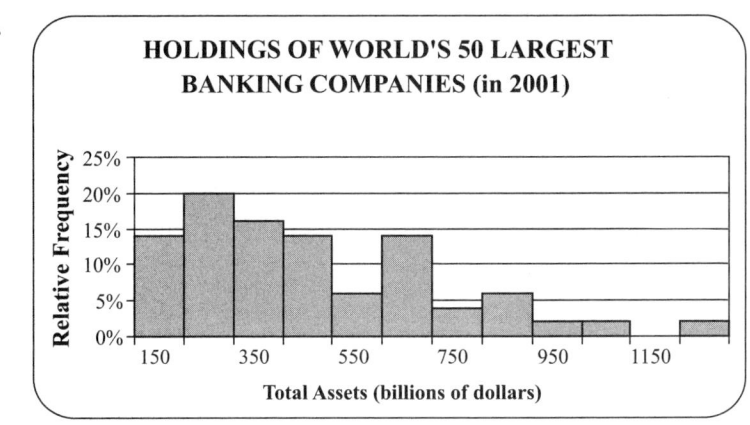

d. 200 to 300 billion dollars

Probability Distributions

Finite Random Variables

Random variables were first introduced in **Expected Value**. Further properties will be discussed in this section.

A **finite random variable** is a random variable that can assume only a finite number of distinct values. For example, finite random variables can represent the number of heads obtained in 20 tosses of a fair coin, the number of cars in an automobile dealership's inventory at the end of the month, the number of students who are logged on to a university's computer network during one hour, etc.

The **probability mass function** (*p.m.f.*) of a finite random variable X is given by

$$f_X(x) = P(X = x).$$

A probability mass function must satisfy the two properties given below.

1. $0 \leq f_X(x) \leq 1$

2. $\sum_{\text{all } x} f_X(x) = 1$

Example 1a: Jack has three pennies, four nickels, two dimes, and one quarter in his pocket. Let X be the denomination (in cents) of a coin selected at random from his pocket.

The *p.m.f.* of X is given below.

x (in cents)	1	5	10	25
$f_X(x)$	0.3	0.4	0.2	0.1

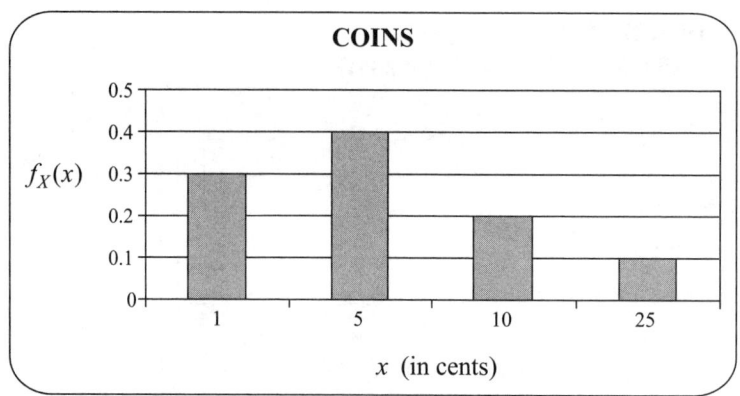

The probability that Jack selects a quarter is given by $P(X = 25) = f_X(25) = 0.1$.
The probability that Jack selects at most five cents is given by

$$P(X \leq 5) = P(X = 1) + P(X = 5) = f_X(1) + f_X(5) = 0.3 + 0.4 = 0.7.$$

The **cumulative distribution function** (*c.d.f.*) of any random variable X is given by
$$F_X(x) = P(X \leq x).$$

The domain of any cumulative distribution function is the set of real numbers, and a *c.d.f.* must satisfy the two properties given below.

1. $0 \leq F_X(x) \leq 1$

2. $F_X(x) \to 0$ as $x \to -\infty$ and $F_X(x) \to 1$ as $x \to \infty$

Example 1b: The *c.d.f.* of X is given by

$$F_X(x) = \begin{cases} 0 & \text{if } x < 1 \\ 0.3 & \text{if } 1 \leq x < 5 \\ 0.7 & \text{if } 5 \leq x < 10 \\ 0.9 & \text{if } 10 \leq x < 25 \\ 1.0 & \text{if } x \geq 25 \end{cases}.$$

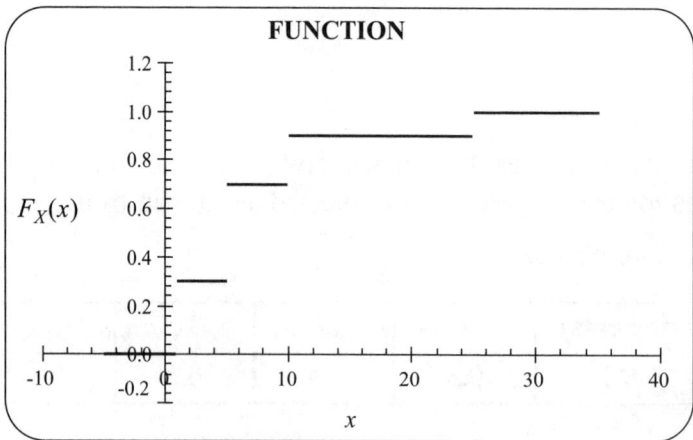

The probability that Jack selects at most five cents is given by

Probability Distributions

$$P(X \leq 5) = F_X(5) = 0.7.$$

Notice that it is easier to use the *c.d.f.* rather than the *p.m.f.* to find this probability.

The expected value or mean of a finite random variable is given by

$$E(X) = \mu_X = \sum_{\text{all } x} x \cdot f_X(x).$$

Example 1c: $E(X) = \mu_X = 1 \cdot 0.3 + 5 \cdot 0.4 + 10 \cdot 0.2 + 25 \cdot 0.1 = 6.8$ cents.

Example 2: The probability mass function of a finite random variable, Y, is given below.

y	0	1	2	3
$f_Y(y)$	0.6141	0.3251	0.0574	0.0034

a. Find $P(Y = 1)$.
b. Find $P(Y \leq 1)$.
c. Find $F_Y(2)$.
d. Find $P(Y \geq 2)$.
e. Find $f_Y(2)$.
f. Find $E(Y)$.

SOLUTION

a. $P(Y = 1) = f_Y(1) = 0.3251$

b. $P(Y \leq 1) = P(Y = 0) + P(Y = 1)$
$= f_Y(0) + f_Y(1)$
$= 0.6141 + 0.3251$
$= 0.9392$

c. $F_Y(2) = P(Y \leq 2)$
$= P(Y = 0) + P(Y = 1) + P(Y = 2)$
$= 0.6141 + 0.3251 + 0.0574$
$= 0.9966$

d. $P(Y \geq 2) = P(Y = 2) + P(Y = 3)$
$= 0.0574 + 0.0034$
$= 0.0608$

or

$P(Y \geq 2) = 1 - P(Y \leq 1)$
$= 1 - 0.9392$
$= 0.0608$

e. $f_Y(2) = 0.0574$

f. $E(Y) = 0 \cdot 0.6141 + 1 \cdot 0.3251 + 2 \cdot 0.0574 + 3 \cdot 0.0034$
    $= 0.4501$

Example 3: The c.d.f. of Y, the number of years among the next five in which a stock index falls, is given below.

$$F_Y(y) = \begin{cases} 0 & \text{if } y < 0 \\ 0.12 & \text{if } 0 \le y < 1 \\ 0.40 & \text{if } 1 \le y < 2 \\ 0.75 & \text{if } 2 \le y < 3 \\ 0.95 & \text{if } 3 \le y < 4 \\ 0.99 & \text{if } 4 \le y < 5 \\ 1 & \text{if } y \ge 5 \end{cases}$$

a. Find the values of the p.m.f. of Y.
b. Create a graph of the p.m.f. of Y.
c. What value of Y is most likely?
d. What value of Y is least likely?
e. Find $P(Y = 4)$.
f. Find $P(Y \le 3)$.

SOLUTION

a.

y	$f_Y(y)$
0	**0.12**
1	0.40 – 0.12 = **0.28**
2	0.75 – 0.40 = **0.35**
3	0.95 – 0.75 = **0.20**
4	0.99 – 0.95 = **0.04**
5	1.00 – 0.99 = **0.01**

b.
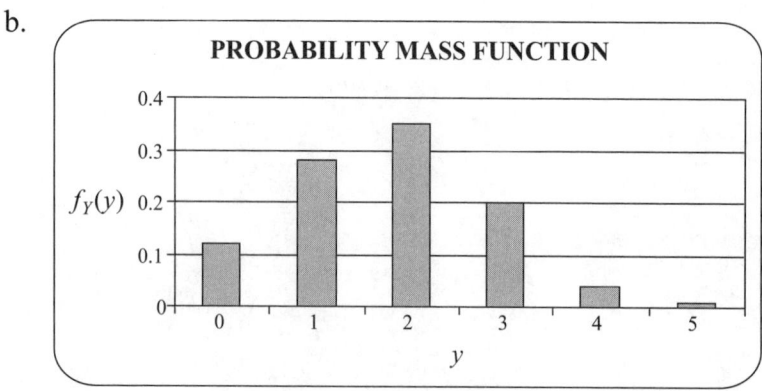

c. The maximum probability is 0.35 for $Y = 2$.
d. The minimum probability is 0.01 for $Y = 5$.
e. $P(Y = 4) = 0.04$.
f. $P(Y \leq 3) = 0.95$.

Binomial Random Variables

A Bernoulli trial is an experiment that has exactly two outcomes, success and failure. A **Bernoulli random variable**, X, assumes the value one for a success and the value zero for a failure. The *p.m.f.* of a Bernoulli random variable X for which the probabilities of success and failure are p and $1 - p$, respectively, is given in the table to the right.

x	$f_X(x)$
1	p
0	$1 - p$

The expected value of a Bernoulli random variable is

$$E(X) = 1 \cdot p + 0 \cdot (1 - p) = p.$$

Example 4: A person goes into an old-fashioned ice cream parlor to get a single scoop of ice cream. There are only three choices: vanilla, chocolate, and strawberry. Let $X = 1$ if the person orders a scoop of vanilla ice cream and let $X = 0$ if the person does not order a scoop of vanilla ice cream. Based on prior sale receipts, the probability of ordering a scoop of vanilla ice cream is 40%. Find the *p.m.f.* of X, compute the expected value of X, and interpret the results.

SOLUTION

The *p.m.f.* of X is given below.

x	$f_X(x)$
1	0.40
0	0.60

$$E(X) = 1 \cdot 0.4 + 0 \cdot (1 - 0.4) = 0.4$$

On average, for a large number of customers, the probability that a customer will order a scoop of vanilla ice cream is 40%.

A **binomial random variable** gives the number of successes in n independent Bernoulli trials, where p is the probability of success on any trial.

Example 5: A recent Gallup Poll of investors' confidence showed that 7 out of 10 American investors believe that now is a good time to invest. Three American investors are selected at random. Let X be the number of those investors who believe that now is a good time to invest. X is a binomial random variable with $n = 3$ and $p = 0.70$. Find the *p.m.f.* and *c.d.f.* of X.

SOLUTION

Let G be the event that an investor believes that now is a good time to invest. Let B be the event that an investor does not believe that now is a good time to invest.

Since 7 out of 10 American investors believe that now is a good time to invest, then $P(G) = 0.70$ and $P(B) = 1 - 0.70 = 0.30$. Tree diagrams are used to find the outcomes in the sample space.

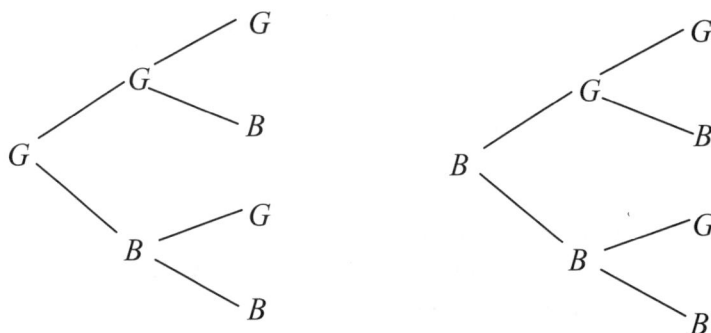

The sample space is given by $S = \{GGG, GGB, GBG, GBB, BBB, BBG, BGB, BGG\}$.

The probability that all three investors believe that now is a good time to invest is given by $P(GGG) = P(G \cap G \cap G)$. Since the opinions of the investors are independent,

$$P(G \cap G \cap G) = P(G) \cdot P(G) \cdot P(G) = (P(G))^3 = (0.70)^3 = 0.343.$$

$$P(X = 3) = P(GGG) = 0.343$$

The probability that the first two investors believe that now is a good time to invest and the third investor does not believe that now is a good time to invest is given by $P(GGB) = P(G \cap G \cap B)$. Since the opinions of the investors are independent,

$$P(G \cap G \cap B) = P(G) \cdot P(G) \cdot P(B) = (P(G))^2 \cdot P(B) = (0.70)^2 \cdot 0.30 = 0.147.$$

The order of the responses does not matter; therefore, $P(GGB) = P(GBG) = P(BGG) = 0.147$. The events GGB, GBG, and BGG are mutually exclusive, so

$$P(X = 2) = P(GGB \cup BGG \cup GBG)$$
$$= P(GGB) + P(BGG) + P(GBG)$$
$$= 0.147 + 0.147 + 0.147$$
$$= 0.441.$$

The probability that the first investor believes that now is a good time to invest and the second and third investors do not believe that now is a good time to invest is given by

$$P(GBB) = P(G \cap B \cap B).$$

Since the opinions of the investors are independent,

$$P(G \cap B \cap B) = P(G) \cdot P(B) \cdot P(B) = P(G) \cdot (P(B))^2 = 0.70 \cdot (0.30)^2 = 0.063.$$

The order of the responses does not matter; therefore, $P(GBB) = P(BBG) = P(BGB) = 0.063$. The events GBB, BBG, and BGB are mutually exclusive, so

$$P(X = 1) = P(GBB \cup BGB \cup BBG)$$
$$= P(GBB) + P(BGB) + P(BBG)$$
$$= 0.063 + 0.063 + 0.063$$
$$= 0.189.$$

The probability that none of the investors believe that now is a good time to invest is given by $P(BBB) = P(B \cap B \cap B)$. Since the opinions of the investors are independent,

$$P(B \cap B \cap B) = P(B) \cdot P(B) \cdot P(B) = (P(B))^3 = (0.30)^3 = 0.027.$$

$$P(X = 0) = P(BBB) = 0.027$$

Recall that the probability mass function (*p.m.f.*) of a finite random variable X is given by

$$f_X(x) = P(X = x),$$

where x is a value of X.

The computations are summarized in the table below.

x	0	1	2	3
$f_X(x)$	0.027	0.189	0.441	0.343

We can use the fact that $\sum_{\text{all } x} f_X(x)$ must be equal to 1 to verify the results.

$$\sum_{\text{all } x} f_X(x) = f_X(0) + f_X(1) + f_X(2) + f_X(3)$$
$$= 0.027 + 0.189 + 0.441 + 0.343$$
$$= 1$$

We can also fill in the table for the *c.d.f.* of X. Recall $F_X(x) = P(X \leq x)$. Therefore,

$$F_X(0) = P(X \leq 0) = P(X = 0) = 0.027$$
$$F_X(1) = P(X \leq 1) = P(X = 0) + P(X = 1) = 0.027 + 0.189 = 0.216$$
$$F_X(2) = P(X \leq 2) = P(X = 0) + P(X = 1) + P(X = 2)$$
$$= 0.027 + 0.189 + 0.441 = 0.657$$
$$F_X(3) = P(X \leq 2) = P(X = 0) + P(X = 1) + P(X = 2) + P(X = 3)$$
$$= 0.027 + 0.189 + 0.441 + 0.343 = 1.$$

x	0	1	2	3
$F_X(x)$	0.027	0.216	0.657	1.000

The expected value of a binomial random variable is given by $E(X) = \mu_X = np$.

The expected value of the random variable defined in **Example 4** is given by $E(X) = 3 \cdot 0.70 = 2.10$.

We can verify the expected value by using the definition of the expected value for any finite random variable.

$$E(X) = \sum_{\text{all } x} x \cdot f_X(x)$$
$$= 0 \cdot 0.027 + 1 \cdot 0.189 + 2 \cdot 0.441 + 3 \cdot 0.343$$
$$= 2.10$$

Example 6: Use the ***BINOMDIST*** function in *Excel* to verify the results found in **Example 5**.

SOLUTION

Start in *Excel* and create the table shown below.

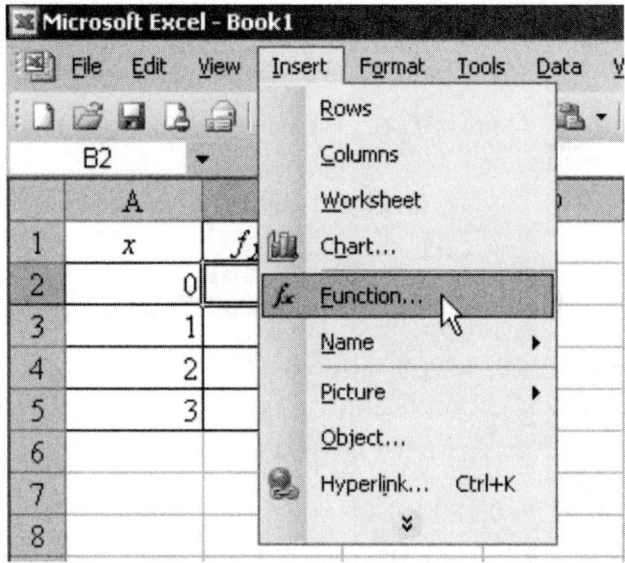

Follow the steps below to compute the values of the *p.m.f.* and *c.d.f.* for the random variable *X*.

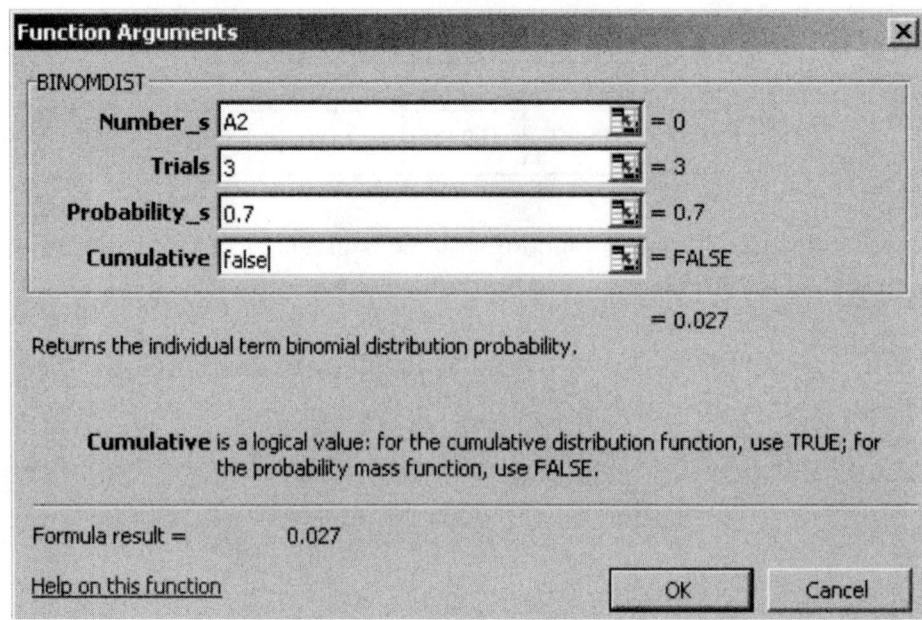

Click OK and the first value of the p.m.f. will appear; $f_X(0) = 0.027$. Copy the formula for the other values of the random variable. To compute the values of the c.d.f., you will follow the same steps as above except for the last input in **BINOMDIST** you will enter "true."

The following table below gives the p.m.f. and c.d.f. for the random variable X in **Example 5**.

x	$f_X(x)$	$F_X(x)$
0	0.027	0.027
1	0.189	0.216
2	0.441	0.657
3	0.343	1

Example 7: A coin that is biased so that the probability of tails is 0.40 is tossed 20 times. Let X be the random variable that represents the number of tails observed. X is a binomial random variable with parameters $n = 20$ and $p = 0.40$. Calculate $P(X = 15)$ and $P(X \leq 10)$.

SOLUTION

The results in the table below were obtained from using **BINOMDIST** with $n = 20$ and $p = 0.40$.

x	$P(X = x)$	$P(X \leq x)$
0	0.000036562	0.000036562
1	0.000487488	0.000524049
2	0.003087423	0.003611472
3	0.012349691	0.015961163
4	0.034990790	0.050951953
5	0.074647020	0.125598973
6	0.124411699	0.250010672
7	0.165882266	0.415892938
8	0.179705788	0.595598725
9	0.159738478	0.755337203
10	0.117141551	0.872478754
11	0.070994879	0.943473633
12	0.035497440	0.978971073
13	0.014563052	0.993534125
14	0.004854351	0.998388475
15	0.001294494	0.999682969
16	0.000269686	0.999952655
17	0.000042304	0.999994959
18	0.000004700	0.999999659
19	0.000000330	0.999999989
20	0.000000011	1.000000000

$P(X = 15) \cong 0.001294494$

$P(X \leq 10) \cong 0.872478754$

Continuous Random Variables

A **continuous random variable** is a random variable that can assume any value in some interval of numbers. For example, continuous random variables can represent the interval of time it takes for a balloon to lose all of its air, the depth of the water in a bucket until it completely evaporates, or the length of time until a candle completely melts, etc.

The **probability density function**, (*p.d.f.*), of a continuous random variable, X, is given by $f_X(x)$. The probability that X is between a and b is equal to the area between the graph of $f_X(x)$ and the horizontal axis over the interval [a, b]. That is,

$$P(a \leq X \leq b) = \text{area under the graph of } f_X(x) \text{ over the interval } [a,b].$$

A *p.d.f.* must satisfy the two properties given below.

1. $f_X(x) \geq 0$ for all x.
2. The total area under the curve must be equal to one.

If X is any continuous random variable, then $P(X = x) = 0$. Why?

Consider T, the time it takes for a balloon to lose all of its air. Let's say that it takes between 0 and 10 seconds for the balloon to completely lose all of its air. The probability that the balloon loses all of its air in 6.53 seconds is given by the area between the graph of $f_T(t)$ over the point 6.53. Since the point 6.53 is equivalent to the interval [6.53, 6.53] and since the width of this interval is zero, the area of interest is zero as well.

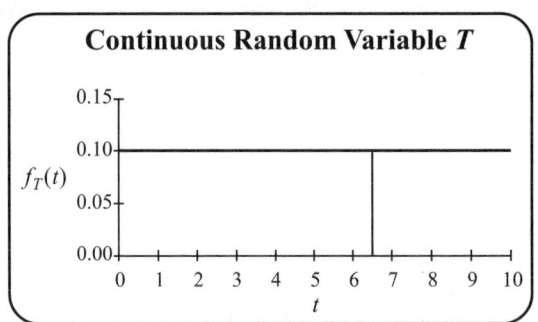

Because the probability of any single value is zero, the events of interest are given by intervals of numbers.

Note that the definition for the *c.d.f.* of a continuous random variable is the same as for a finite random variable — $F_X(x) = P(X \leq x)$. The probability that X is between a and b is given by the difference between $F_X(b)$ and $F_X(a)$. That is, $P(a \leq X \leq b) = F_X(b) - F_X(a)$.

Example 8: The formulas for and graphs of the *p.d.f.* and *c.d.f.* of a continuous random variable, X, are given below.

$$f_X(x) = \begin{cases} 0 & \text{if } x < 0 \\ 0.005x & \text{if } 0 \leq x \leq 20 \\ 0 & \text{if } x > 20 \end{cases} \qquad F_X(x) = \begin{cases} 0 & \text{if } x < 0 \\ 0.0025x^2 & \text{if } 0 \leq x < 20 \\ 1 & \text{if } x \geq 20 \end{cases}$$

Probability Distributions

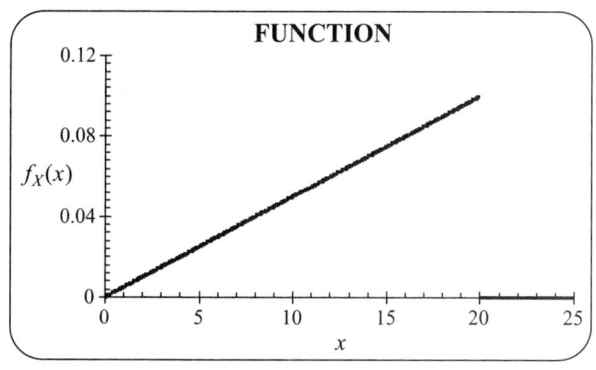 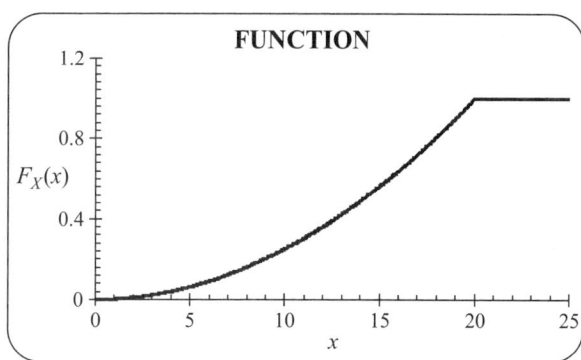

a. Use the *p.d.f.* to compute $P(X \leq 15)$.

b. Use the formula for the *c.d.f.* to compute $P(X \leq 15)$.

SOLUTION

a. $P(X \leq 15)$ is given by the area between the graph of the *p.d.f.* and the horizontal axis over the interval $[0, 15]$.

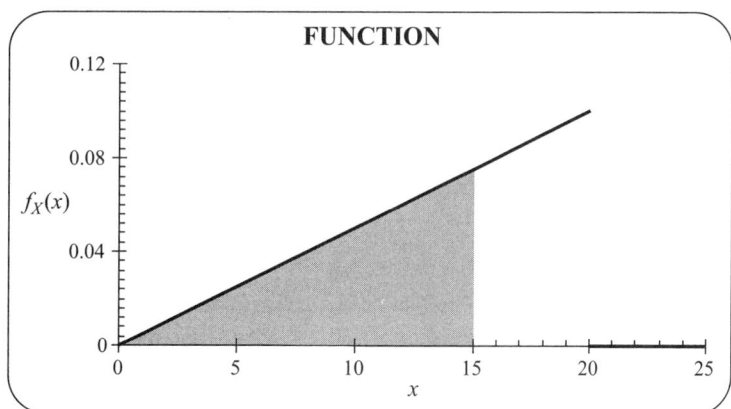

This is the area of a triangle with base $15 - 0 = 15$ and height $f_X(15) = 0.005 \cdot 15 = 0.075$. Therefore, $P(X \leq 15) = 1/2 \cdot 15 \cdot 0.075 = 0.5625$.

b. $P(X \leq 15) = F_X(15) = 0.0025 \cdot 15^2 = 0.5625$

Example 9: The formulas for and graphs of the *p.d.f.* and *c.d.f.* of a continuous random variable, X, are given below.

$$f_X(x) = \begin{cases} 0 & \text{if } x < 0 \\ 0.25e^{-x/2} & \text{if } x \geq 0 \end{cases} \qquad F_X(x) = \begin{cases} 0 & \text{if } x < 0 \\ 1 - e^{-x/2} - 0.5xe^{-x/2} & \text{if } x \geq 0 \end{cases}$$

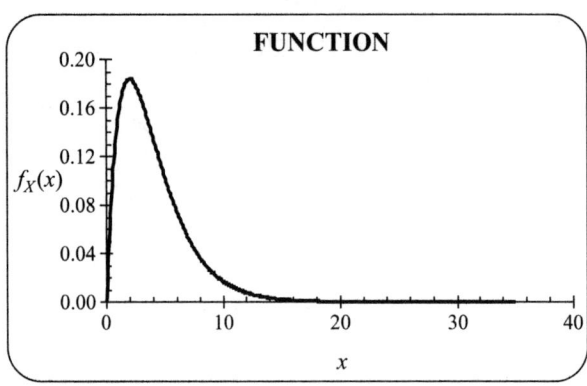 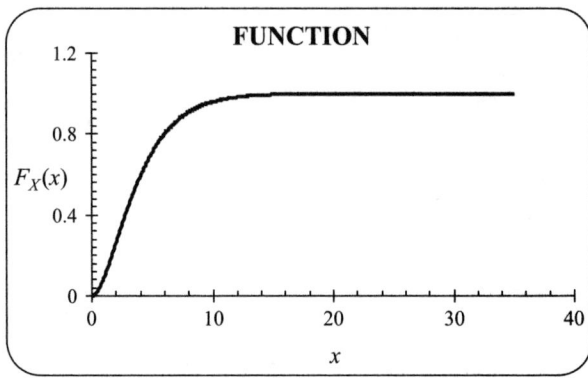

a. Shade the region of the graph of the p.d.f. that corresponds to $P(2 \leq X \leq 4)$.

b. Use the formula for the c.d.f. to compute $P(2 \leq X \leq 4)$.

SOLUTION

a. The area between the graph of the p.d.f. and the horizontal axis over the interval [2, 4] corresponds to $P(2 \leq X \leq 4)$.

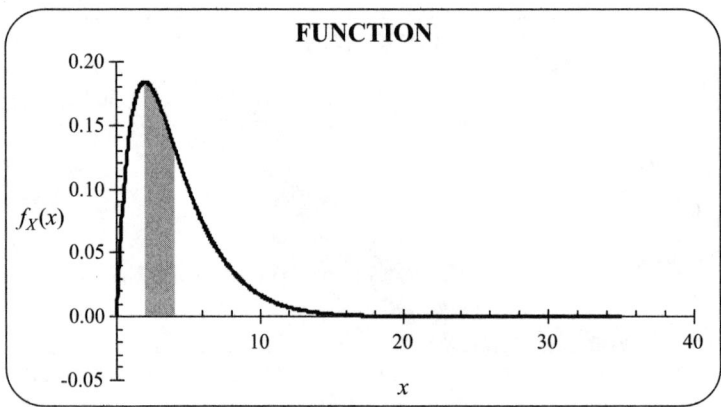

b. $P(2 \leq X \leq 4) = P(X \leq 4) - P(X < 2)$
$= P(X \leq 4) - P(X \leq 2)$
$= F_X(4) - F_X(2)$
$= (1 - e^{-4/2} - 0.5 \cdot 4 e^{-4/2}) - (1 - e^{-2/2} - 0.5 \cdot 2 e^{-2/2})$
$\approx 0.5940 - 0.2642$
$= 0.3298$

Uniform Random Variables

A continuous random variable defined on an interval [a, b] has a **uniform distribution** if every subinterval of [a, b] having the same length has the same probability. The p.d.f. of a random variable that has a uniform distribution on the interval [a, b] is given by

Probability Distributions

$$f_X(x) = \begin{cases} 0 & \text{if } x < a \\ \dfrac{1}{b-a} & \text{if } a \leq x \leq b \\ 0 & \text{if } x > b \end{cases}.$$

We now show that this *p.d.f.* satisfies the properties that must be satisfied by any *p.d.f.*

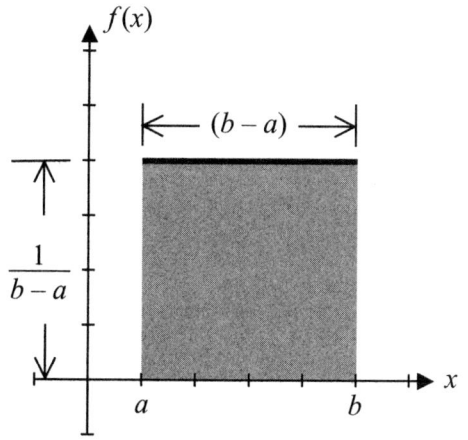

Note that:

1. $f_X(x) \geq 0$ for all x.
   The graph is strictly on or above the horizontal axis.

2. The total area under the curve is one.
   Area under the graph = width · height
   $$= (b-a) \cdot \frac{1}{b-1}$$
   $$= 1$$

Example 10a: The graph of the *p.d.f.* of a random variable that has a uniform distribution on the interval [2, 6] is given below. Find the equation of the *p.d.f.*

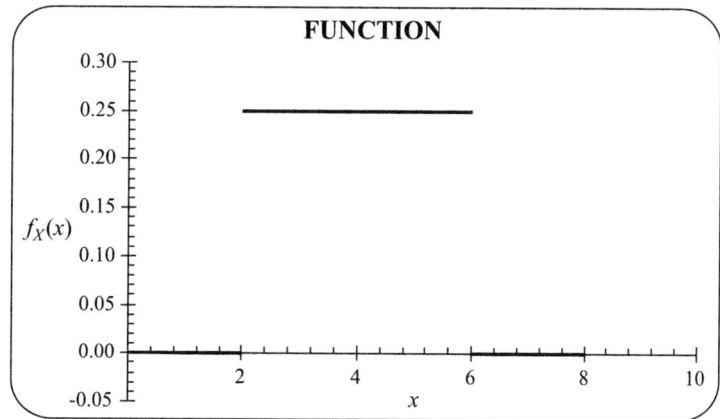

SOLUTION

$$(b-a) = (6-2) = 4$$

$$f_X(x) = \begin{cases} 0 & \text{if } x < 2 \\ \dfrac{1}{4} & \text{if } 2 \leq x \leq 6 \\ 0 & \text{if } x > 6 \end{cases}$$

We can derive the *c.d.f.* of a random variable that has a uniform distribution on the interval $[a, b]$. The cumulative probability is zero for x less than a (the smallest possible value of X) and one for x greater than b (the largest possible value of X). Furthermore, the cumulative probability function is a linear function on the interval from a to b. So now we can find the equation of that line.

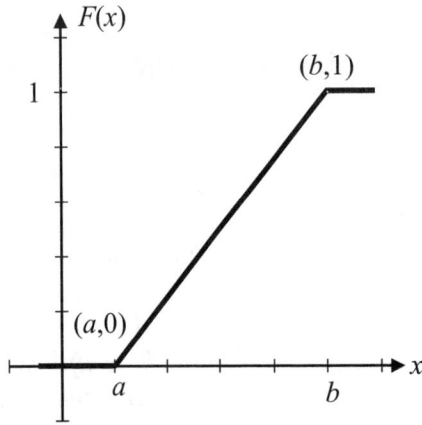

The slope of the line is given by

$$m = \frac{\Delta y}{\Delta x} = \frac{y_2 - y_1}{x_2 - x_1} = \frac{1-0}{b-a} = \frac{1}{b-a}.$$

Using the point-slope form of the equation of a line, $y - y_1 = m(x - x_1)$, we have

$$y - 0 = \frac{1}{b-a} \cdot (x - a) \quad \text{or} \quad y = \frac{x-a}{b-a}.$$

The *c.d.f.* of a random variable that has a uniform distribution on the interval $[a, b]$ is given by

$$F_X(x) = \begin{cases} 0 & \text{if } x < a \\ \dfrac{x-a}{b-a} & \text{if } a \leq x < b \\ 1 & \text{if } x \geq b. \end{cases}$$

Example 10b: The graph of the *c.d.f.* of a random variable that has a uniform distribution on the interval $[2, 6]$ is given below. Find the equation of the *c.d.f.*

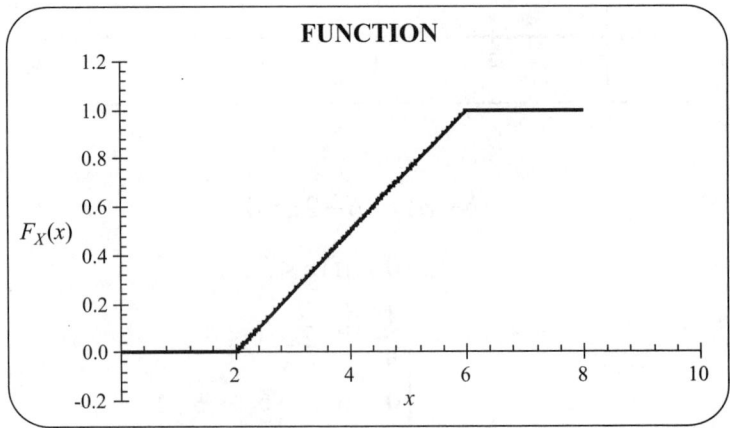

Probability Distributions

SOLUTION

$$(b-a) = (6-2) = 4$$

$$F_X(x) = \begin{cases} 0 & \text{if } x < 2 \\ \dfrac{x-2}{4} & \text{if } 2 \leq x < 6 \\ 1 & \text{if } x \geq 6 \end{cases}$$

Example 10c: The graphs of the *p.d.f.* and *c.d.f.* of a random variable that has a uniform distribution on the interval [2,6] are given below.

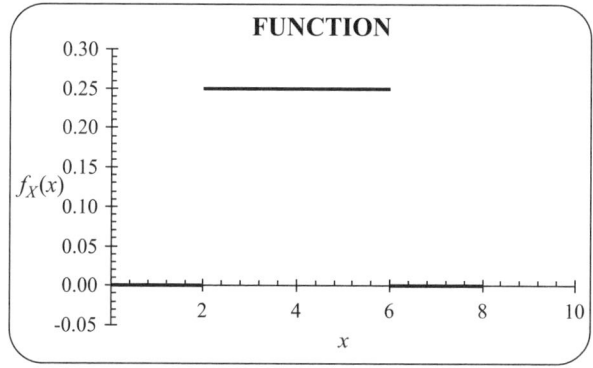
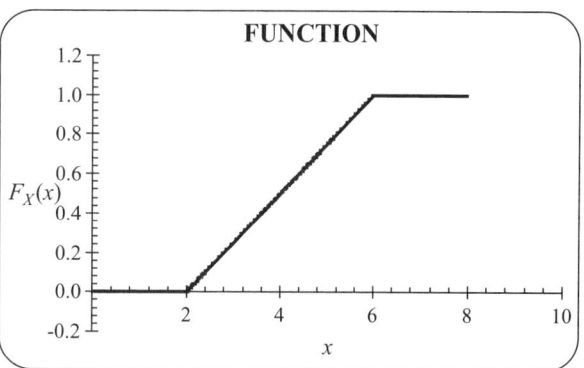

a. Use the equation of the *c.d.f.* to compute $P(X \leq 5)$.
b. Use the graph of the *p.d.f.* to compute $P(3 \leq X \leq 5)$.
c. Use the graph of the *c.d.f.* to approximate $P(3 \leq X \leq 5)$.
d. Use the equation of the *c.d.f* to compute $P(3 \leq X \leq 5)$.
e. Compute $P(X = 3.5)$.

SOLUTION

a. $P(X \leq 5) = F_X(5) = \dfrac{5-2}{4} = \dfrac{3}{4} = 0.75$

b.

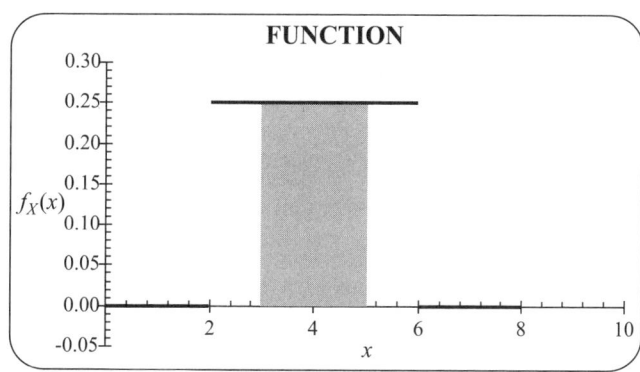

$P(3 \leq X \leq 5) = \text{area of shaded region} = \text{width} \cdot \text{height} = (5-3) \cdot \dfrac{1}{4} = 0.5$

c.

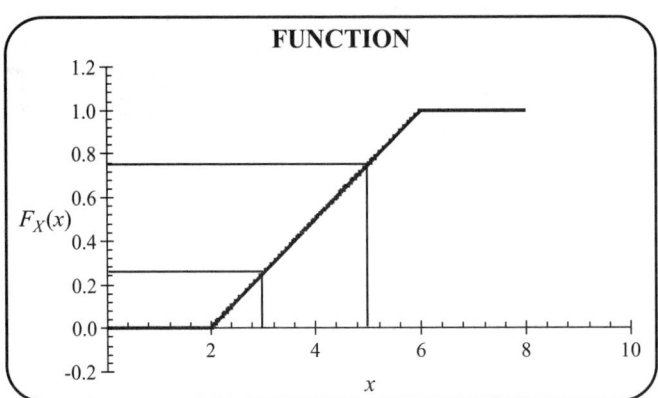

$$P(3 \le X \le 5) = F_X(5) - F_X(3) \cong 0.72 - 0.23 = 0.49$$

d. $P(3 \le X \le 5) = F_X(5) - F_X(3) = \dfrac{5-2}{4} - \dfrac{3-2}{4} = \dfrac{3}{4} - \dfrac{1}{4} = \dfrac{2}{4} = 0.50$

e. $P(X = 3.5) = 0$

The expected value of a random variable that has a uniform distribution on the interval [a, b] is given by

$$E(X) = \mu_X = \frac{a+b}{2}.$$

The expected value can be viewed as the balancing point for the area under the curve between a and b.

Example 10d: Compute the expected value of X, a random variable that has a uniform distribution on the interval [2,6].

SOLUTION

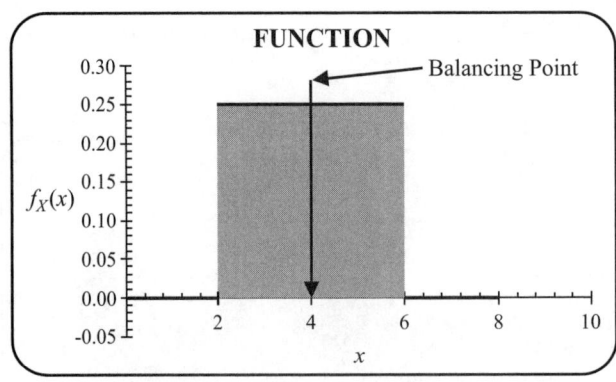

$$E(X) = \mu_X = \frac{2+6}{2} = 4$$

Example 11: The length of time (in minutes), T, required for an exam has a uniform distribution on the interval [40, 50].

a. Graph the *p.d.f.* of T on the interval $[0, 60]$.
b. Find a formula for the *p.d.f.* of T.
c. Graph the *c.d.f.* of T on the interval $[0, 60]$.
d. Find a formula for the *c.d.f.* of T.
e. Find the probability that more than 48 minutes is required for the exam.
f. Find the probability that less than 48 minutes is required for the exam.
g. Find the average length of time required for the exam.

SOLUTION

a.

[Graph of $f_T(t)$ vs t on $[0, 60]$, showing a constant value of 0.10 on $[40, 50]$ and 0 elsewhere.]

b. $f_T(t) = \begin{cases} 0 & \text{if } t < 40 \\ \dfrac{1}{10} & \text{if } 40 \leq t \leq 50 \\ 0 & \text{if } t > 50 \end{cases}$

c.

[Graph of $F_T(t)$ vs t on $[0, 60]$, showing 0 on $[0, 40]$, linear increase from 0 to 1 on $[40, 50]$, and constant 1 on $[50, 60]$.]

d. $F_T(t) = \begin{cases} 0 & \text{if } t < 40 \\ \dfrac{t-40}{10} & \text{if } 40 \leq t < 50 \\ 0 & \text{if } t \geq 50 \end{cases}$

e. $P(T > 48) = 1 - P(T \leq 48) = 1 - F_T(48) = 1 - \dfrac{48-40}{10} = 0.20$

Note: The *p.d.f.* can also be used to answer this question.

$P(T > 48)$ = area under curve over the interval [48, 50]

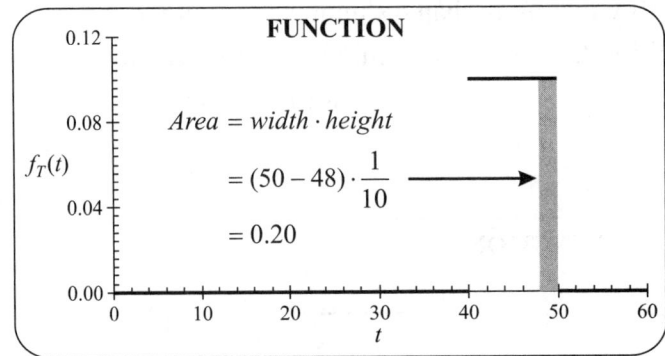

f. $P(T < 48) = P(T \leq 48) = F_T(48) = \dfrac{48-40}{10} = 0.80$

Note: The *p.d.f.* can also be used to answer this question.

$P(T < 48)$ = area under curve over the interval [40, 48]

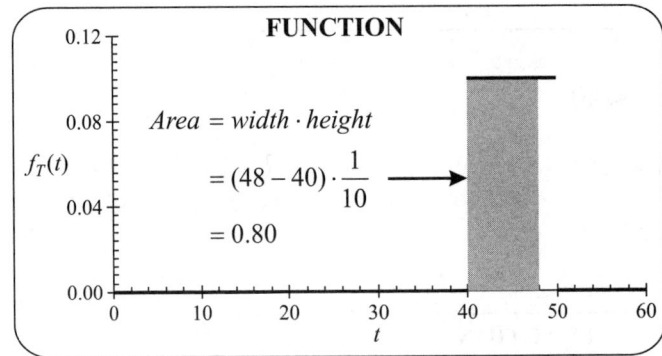

g. $E(T) = \dfrac{40+50}{2} = 45$ minutes

Note: This is the balancing point for the area under the graph of the *p.d.f.*, which is shown graphically below.

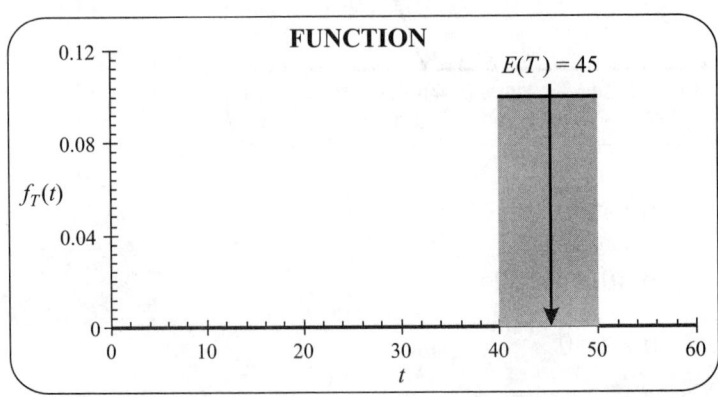

Probability Distributions

Example 12: A bus arrives at a bus stop every ten minutes. If T is the waiting time (in minutes) until the next bus, then T has a uniform distribution on the interval [0, 10].

a. Graph the *p.d.f.* of T on the given interval.
b. Find a formula for the *p.d.f.* of T.
c. Graph the *c.d.f.* of T on the given interval.
d. Find a formula for the *c.d.f.* of T.
e. Find the probability that the waiting time for the bus will be more than six minutes.
f. Find the probability that the waiting time for the bus will be less than three minutes.
g. Find the average waiting time for the bus.

SOLUTION

a.
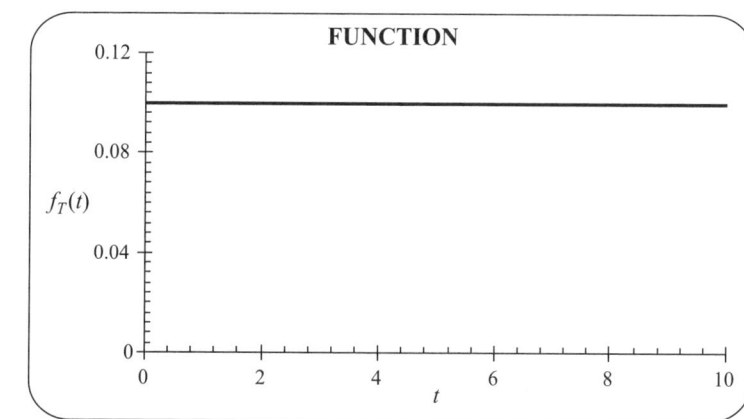

b. $f_T(t) = \begin{cases} 0 & \text{if } t < 0 \\ \dfrac{1}{10} & \text{if } 0 \leq t \leq 10 \\ 0 & \text{if } t > 10 \end{cases}$

c.
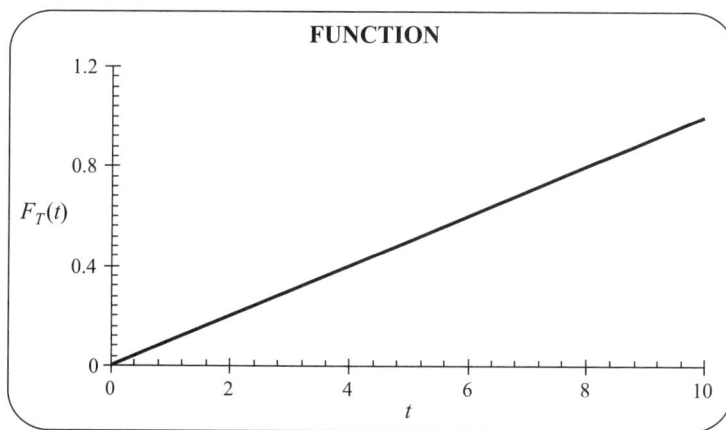

d. $F_T(t) = \begin{cases} 0 & \text{if } t < 0 \\ \dfrac{t}{10} & \text{if } 0 \le t < 10 \\ 1 & \text{if } t \ge 10 \end{cases}$

e. $P(T > 6) = 1 - P(T \le 6) = 1 - F_T(6) = 1 - \dfrac{6}{10} = 0.40$

f. $P(T < 3) = P(T \le 3) = F_T(3) = \dfrac{3}{10} = 0.30$

g. $E(T) = \dfrac{0 + 10}{2} = 5$ minutes

Exponential Random Variables

Exponential random variables are used to model the length of time between consecutive occurrences of some event in a fixed unit of space or time.

The *p.d.f.* of an exponential random variable X with parameter α is given by

$$f_X(x) = \begin{cases} 0 & \text{if } x < 0 \\ \dfrac{1}{\alpha} e^{-x/\alpha} & \text{if } x \ge 0, \end{cases}$$

and the *c.d.f.* of X is given by

$$F_X(x) = \begin{cases} 0 & \text{if } x < 0 \\ 1 - e^{-x/\alpha} & \text{if } x \ge 0. \end{cases}$$

The expected value of an exponential random variable with parameter α is given by

$$E(X) = \mu = \alpha.$$

Example 13: Let X be the continuous random variable that gives the time (in minutes) between departures of shuttle busses from the Disneyland parking lot. It is known that there are 0.25 departures per minute.

a. Find a formula for the *p.d.f.* of X.
b. Find a formula for the *c.d.f.* of X.
c. Find the expected value of X.
d. Compute $P(2 \le X \le 4)$ using the *c.d.f.*

SOLUTION

a. $\dfrac{1}{\alpha \text{ minutes}} = \dfrac{0.25 \text{ departures}}{1 \text{ minute}}$

$$0.25\alpha = 1$$
$$\alpha = \frac{1}{0.25} = 4$$

$$f_X(x) = \begin{cases} 0 & \text{if } x < 0 \\ \frac{1}{4}e^{-x/4} & \text{if } x \geq 0 \end{cases}$$

b. $F_X(x) = \begin{cases} 0 & \text{if } x < 0 \\ 1 - e^{-x/4} & \text{if } x \geq 0 \end{cases}$

c. $E(X) = \alpha = 4$ minutes

d. $P(2 \leq X \leq 4) = F_X(4) - F_X(2) = (1 - e^{-4/4}) - (1 - e^{-2/4}) \cong 0.2387$

Example 14: Let X be the continuous random variable that gives the time (in minutes) between the arrivals/departures of planes at the Phoenix Sky Harbor Airport. There are 8.3 planes arriving or departing every ten minutes.
a. Find the *p.d.f.* of X.
b. Find the *c.d.f.* of X.
c. Use the *c.d.f.* to find the probability that the time between the arrivals/departures at the Phoenix Sky Harbor Airport is at most three minutes.
d. Use the *c.d.f.* to find the probability that the time between the arrivals/departures at the Phoenix Sky Harbor Airport is between one and three minutes.

SOLUTION

a. $\dfrac{1 \text{ arrival/departure}}{\alpha \text{ minutes}} = \dfrac{8.3 \text{ arrivals/departures}}{10 \text{ minutes}}$

$$8.3\alpha = 1 \cdot 10$$
$$\alpha = \frac{10}{8.3} \cong 1.2$$

$$f_X(x) = \begin{cases} 0 & \text{if } x < 0 \\ \dfrac{1}{1.2}e^{-x/1.2} & \text{if } x \geq 0 \end{cases}$$

b. $F_X(x) = \begin{cases} 0 & \text{if } x < 0 \\ 1 - e^{-x/1.2} & \text{if } x \geq 0 \end{cases}$

c. $P(X \leq 3) = F_X(3) = 1 - e^{-3/1.2} \cong 0.9179$

d. $P(1 \leq X \leq 3) = F_X(3) - F_X(1)$
$$= (1 - e^{-3/1.2}) - (1 - e^{-1/1.2})$$
$$\cong 0.9179 - 0.5654 = 0.3525$$

Exercises

1. Two fair tetrahedral (4-sided) dice are rolled.

   a. Find the sample space.

   b. Let E be the event that the sum of the numbers is even. List the outcomes in E.

   c. Compute $P(E)$.

   d. Let F be the event that the sum of the numbers is even and less than 5. Compute $P(F)$.

   e. Let X be the finite random variable that gives the sum of the numbers on the faces of the dice. List the value of X for each outcome.

   f. Compute $P(X = 4)$.

   g. Compute $P(X \leq 4)$.

   h. Compute $P(X > 4)$.

SOLUTION

a. $S = \begin{Bmatrix} (1,1) & (1,2) & (1,3) & (1,4) \\ (2,1) & (2,2) & (2,3) & (2,4) \\ (3,1) & (3,2) & (3,3) & (3,4) \\ (4,1) & (4,2) & (4,3) & (4,4) \end{Bmatrix}$

b. $E = \begin{Bmatrix} (1,1) & (1,3) \\ (2,2) & (2,4) \\ (3,1) & (3,3) \\ (4,2) & (4,4) \end{Bmatrix}$

c. $P(E) = \dfrac{8}{16} = 0.5$

d. $P(F) = \dfrac{4}{16} = 0.25$

e.

Outcome	Value of X	Outcome	Value of X
(1,1)	2	(3,1)	4
(1,2)	3	(3,2)	5
(1,3)	4	(3,3)	6
(1,4)	5	(3,4)	7
(2,1)	3	(4,1)	5
(2,2)	4	(4,2)	6
(2,3)	5	(4,3)	7
(2,4)	6	(4,4)	8

f. $P(X = 4) = \dfrac{3}{16} = 0.1875$

g. $P(X \leq 4) = \dfrac{6}{16} = 0.375$

h. $P(X > 4) = \dfrac{10}{16} = 0.625$

2. The p.m.f. of R, the daily demand for a particular model of notebook computer, is given below.

r	0	1	2	3	4	5
$f_R(r)$	0.10	0.40	0.20	0.15	0.10	0.05

a. Find $P(R = 1)$.
b. Find $F_R(1)$.
c. Find the probability that the demand for this particular model of notebook computer is exactly three.
d. Find the probability that the demand for this particular model of notebook is at least three.
e. Find the probability that the demand for this particular model of notebook computer is at most three.
f. Find μ_R.

SOLUTION

a. $P(R = 1) = f_R(1) = 0.40$

b. $F_R(1) = P(R \le 1) = P(R = 0) + P(R = 1) = 0.10 + 0.40 = 0.50$

c. $P(R = 3) = f_R(3) = 0.15$

d. $P(R \ge 3) = P(R = 3) + P(R = 4) + P(R = 5) = 0.15 + 0.10 + 0.05 = 0.30$

e. $P(R \le 3) = P(R = 0) + P(R = 1) + P(R = 2) + P(R = 3)$
$= 0.10 + 0.40 + 0.20 + 0.15$
$= 0.85$

f. $\mu_R = \sum_{\text{all } r} r \cdot f_R(r) = 0 \cdot 0.10 + 1 \cdot 0.40 + 2 \cdot 0.20 + 3 \cdot 0.15 + 4 \cdot 0.10 + 5 \cdot 0.05 = 1.9$

3. Let X be the profit (in thousands of dollars) on an investment in a bookstore. The p.m.f of X is given below.

x	200	40	−10	−125
$f_X(x)$	0.2	0.1	0.4	0.3

a. What is the probability that the profit on an investment in a bookstore is at least $20,000?
b. Find $E(X)$.
c. Give the practical interpretation of the answer found in Part b.

SOLUTION

a. $P(X \ge 20) = P(X = 200) + P(X = 40) = 0.2 + 0.1 = 0.3$

b. $E(X) = 200(0.2) + 40(0.1) + (−10)(0.4) + (−125)(0.3) = 2.5$ thousand dollars

c. The average profit for a large number of similar investments will be approximately 2.5 thousand dollars.

4. The c.d.f. of a finite random variable, X, is given below.

$$F_X(x) = \begin{cases} 0 & \text{if } x < 5 \\ 0.2 & \text{if } 5 \leq x < 10 \\ 0.6 & \text{if } 10 \leq x < 20 \\ 0.9 & \text{if } 20 \leq x < 40 \\ 1 & \text{if } x \geq 40 \end{cases}$$

a. Find the values of the p.m.f. of X.
b. What value of X is most likely?
c. What value of X is least likely?
d. Find $P(X = 20)$.
e. Find $P(X \leq 10)$.
f. Find $E(X)$.

SOLUTION

a.

x	5	10	20	40
$f_X(x)$	0.2	0.4	0.3	0.1

b. 10
c. 40
d. $P(X = 20) = 0.3$
e. $P(X \leq 10) = 0.6$
f. $E(X) = 15$

5. A local computer store, DCS, was recently sold to an entrepreneur, Mr. Ron Flanders. Mr. Flanders would like to know what type of computer his customers own so that he can maintain an inventory that suits their needs. He is conducting a survey and randomly selects four customers per hour and asks if they own a PC or Mac. Past records indicate that 30% of DCS customers own a Mac. Let X be the number of customers in a sample size of four who own a Mac. The p.m.f. of X is given below.

x	$f_X(x)$
0	0.2401
1	0.4116
2	0.2646
3	0.0756
4	0.0081

a. Find the probability that three or more of the surveyed customers own a Mac.

b. Find the probability that none of the surveyed customers own a Mac.

c. Find the probability that only one of the surveyed customers owns a PC.

d. If the computer store expects to survey 10 groups of four customers each day, on average how many of the customers surveyed in one day are expected to own a Mac?

SOLUTION

a. $P(X \geq 3) = P(X = 3) + P(X = 4) = 0.0756 + 0.0081 = 0.0837$

b. $P(X = 0) = 0.2401$

c. $P(X = 3) = 0.0756$

d. If the computer store expects to survey 10 groups of four customers each day, then the average number of customers surveyed in one day who are expected to own a Mac is given by $n \cdot p \cdot$ number of groups of four customers or $4 \cdot 0.30 \cdot 10 = 12$.

6. An oil exploration company finds oil in approximately 5% of the test wells that it drills. Suppose that the company has plans to drill four new test wells. Then Y, the number of those wells that yield oil, is a binomial random variable with parameters $n = 4$ and $p = 0.05$.

   a. Use the *Excel* function **BINOMDIST** to fill in the table.

   b. Find $P(Y = 3)$.

   c. Find $P(Y \leq 3)$.

   d. Find $P(Y > 3)$.

   e. Compute μ_Y.

y	$f_Y(y)$	$F_Y(y)$

SOLUTION

a. See the table to the right.

b. Find $P(Y = 3) \cong 0.00048$

c. $P(Y \leq 3) \cong 0.99999$

d. $P(Y > 3) = 1 - 0.99999 \cong 0.00001$

e. $\mu_Y = np = 4(0.05) = 0.2$ of a well

y	$f_Y(y)$	$F_Y(y)$
0	0.81451	0.81451
1	0.17148	0.98598
2	0.01354	0.99952
3	0.00048	0.99999
4	0.00001	1.00000

7. Students go to a hotdog stand in groups of three. Each student chooses either a hotdog with chili or a hotdog without chili. The probability that a student chooses a chilidog is 0.75. Let X be the number of students in a group of three who buy a chilidog.

   a. Fill in the table to the right.

   b. Find the probability that two or more chilidogs are sold to a group of three students.

   c. What is the probability that exactly two students don't buy chilidogs?

x	$f_X(x)$	$F_X(x)$
	0.0156	
	0.1406	
	0.4219	

d. In words describe the meaning of $F_X(1)$ and calculate its value.
e. In words describe the meaning of $f_X(3)$ and calculate its value.
f. If the food stand expects 50 groups of three for lunch, how many chilidogs can they expect to sell during an average lunch period?

SOLUTION

a.

x	$f_X(x)$	$F_X(x)$
0	0.0156	0.0156
1	0.1406	0.1562
2	0.4219	0.5781
3	0.4219	1.0000

b. Using $F_X(x)$, $P(X \geq 2) = 1 - P(X \leq 1) = 1 - 0.1562 = 0.8438$.
   Using $f_X(x)$, $P(X \geq 2) = P(X = 2) + P(X = 3) = 0.4219 + 0.4219 = 0.8438$.

c. $P(X = 1) = 0.1406$

d. The probability that at most one student in a group of three buys a chilidog; $F_X(1) = 0.1562$.

e. The probability that all of the students in a group of three buy chilidogs; $f_X(3) = 0.4219$.

f. They can expect to sell $np \cdot$ number of groups of three students or $3 \cdot 0.75 \cdot 50 = 112.5$ chilidogs during an average lunch period.

8. A recent Gallup poll showed that 76% of Americans have at least one credit card. Let T be the number of Americans in a sample of size 20 who have at least one credit card.
   a. Use the **BINOMDIST** function in *Excel* to compute $P(T \leq 16)$.
   b. Compute μ_T.

SOLUTION

a. 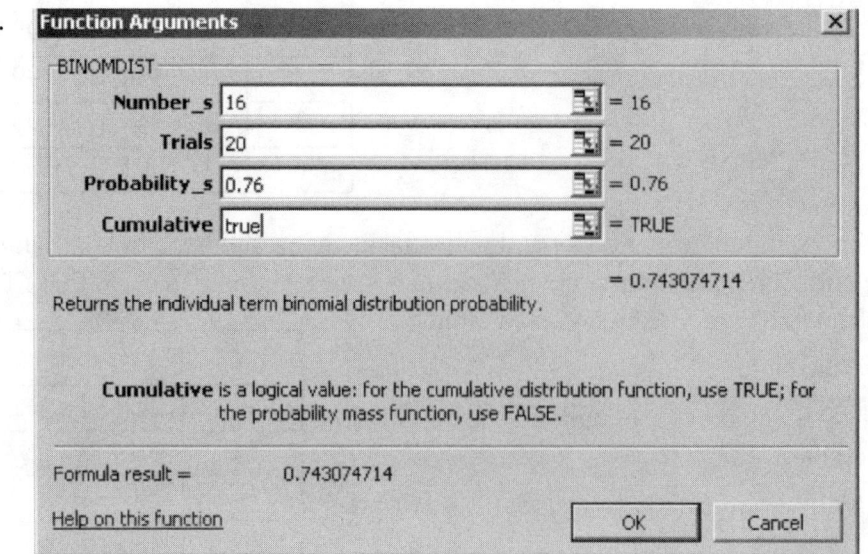 $P(T \leq 16) \cong 0.7431$

b. $\mu_T = np = 20 \cdot 0.76 \cong 15.2$

9. A recent survey showed that 35% of junior college students own a vehicle. Let Y be the event that a junior college student owns a vehicle and let N be the event that a junior college student does not own a vehicle. Let V be the number of junior college students in a sample size of three who own a vehicle. Fill in the table to the right **without** using the **BINOMDIST** function in *Excel*.

v	$f_v(V)$	$F_V(v)$

SOLUTION

Find the sample space, S.

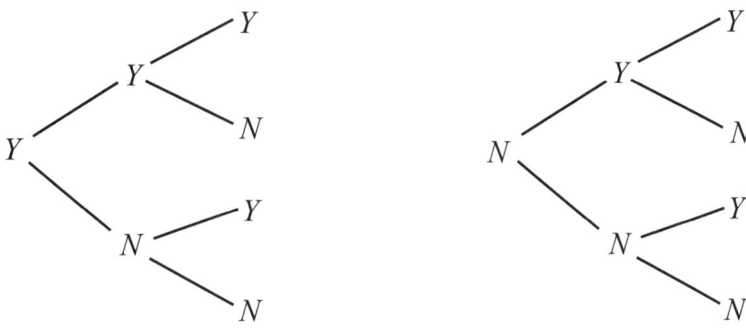

$$S = \{YYY, YYN, YNY, YNN, NYY, NYN, NNY, NNN\}$$
$$P(Y) = 0.35 \quad \text{and} \quad P(N) = 1 - 0.35 = 0.65$$

The probability that all three junior college students own a vehicle is given by

$$P(YYY) = P(Y \cap Y \cap Y) = P(Y) \cdot P(Y) \cdot P(Y) = (P(Y))^3 = 0.35^3 \cong 0.0429.$$
$$P(V = 3) = P(YYY) \cong 0.0429$$

The probability that the first two junior college students own a vehicle and the third student does not own a vehicle is given by

$$P(YYN) = P(Y \cap Y \cap N) = P(Y) \cdot P(Y) \cdot P(N) = (P(Y))^2 \cdot P(N) = 0.35^2 \cdot 0.65 \cong 0.0796.$$
$$P(YYN) = P(YNY) = P(NYY) \cong 0.0796$$

$$P(V = 2) = P(YYN \cup YNY \cup NYY) = P(YYN) + P(YNY) + P(NYY) \cong 3 \cdot 0.0796 = 0.2388$$

The probability that the first junior college student owns a vehicle and the second and third students do not own a vehicle is given by

$$P(YNN) = P(Y \cap N \cap N) = P(Y) \cdot P(N) \cdot P(N) = P(Y) \cdot (P(N))^2 = 0.35 \cdot 0.65^2 \cong 0.1479.$$
$$P(YNN) = P(NYN) = P(NNY) \cong 0.1479$$

$$P(V = 1) = P(YNN \cup NYN \cup NNY) = P(YNN) + P(NYN) + P(NNY) \cong 3 \cdot 0.1479 = 0.4437$$

The probability that none of the junior college students own a vehicle is given by

$$P(NNN) = P(N \cap N \cap N) = P(N) \cdot P(N) \cdot P(N) = (P(N))^3 = (0.65)^3 \cong 0.2746.$$
$$P(V = 0) = P(NNN) \cong 0.2746$$

The *p.m.f.* of a finite random variable X is given by $f_X(x) = P(X = x)$, therefore the results from the previous page can be used to fill out the table for the *p.m.f.* The *c.d.f.* of a finite random variable X is given by $F_X(x) = P(X \leq x)$. This formula is used in the following computations.

$$F_V(0) = 0.2746$$
$$F_V(1) = 0.2746 + 0.4437 = 0.7183$$
$$F_V(2) = 0.2746 + 0.4437 + 0.2388 = 0.9571$$
$$F_V(3) = 0.2746 + 0.4437 + 0.2388 + 0.0429 = 1.0000$$

Now the last column in the table can be completed.

v	$f_V(v)$	$F_V(v)$
0	0.2746	0.2746
1	0.4437	0.7183
2	0.2388	0.9571
3	0.0429	1.0000

Are we sure that the calculations are correct? The **BINOMDIST** function in *Excel* is used to verify the results.

0	0.2746	0.2746
1	0.4437	0.7183
2	0.2388	0.9571
3	0.0429	1.0000

10. A recent study showed that one in four college students is age 30 or older. Let W be the random variable that corresponds to the number of college students in a sample of size ten who are age 30 or older.

   a. Use the **BINOMDIST** function in *Excel* to find the *p.m.f.* and *c.d.f.* of W.
   b. Use the *p.m.f.* to find the probability that exactly six students in a sample of size ten are age 30 or older.
   c. Use the *c.d.f.* to find the probability that at most six students in a sample of size ten are age 30 or older.
   d. Use the *p.m.f.* to find the probability that at least seven students in a sample of size ten are age 30 or older.
   e. Use the *c.d.f.* to answer the question in Part d.
   f. Find μ_W using *Excel*.
   g. Find μ_W using the fact that $\mu_W = np$.

Probability Distributions

SOLUTION

a.

w	$f_W(w)$	$F_W(w)$
0	0.056314	0.056314
1	0.187712	0.244025
2	0.281568	0.525593
3	0.250282	0.775875
4	0.145998	0.921873
5	0.058399	0.980272
6	0.016222	0.996494
7	0.003090	0.999584
8	0.000386	0.999970
9	0.000029	0.999999
10	0.000001	1.000000

b. $P(W = 6) = f_W(6) = 0.016222$

c. $P(W \leq 6) = F_W(6) = 0.996494$

d. $P(W \geq 7) = P(W = 7) + P(W = 8) + P(W = 9) + P(W = 10)$
$= 0.003090 + 0.000386 + 0.000029 + 0.000001$
$= 0.003506$

e. $P(W \geq 7) = 1 - P(W \leq 6) = 1 - F_W(6) = 1 - 0.996494 = 0.003506$

f. $\mu_W = \sum_{\text{all } w} w \cdot f_W(w) = 2.5$ students

w	$f_W(w)$	$F_W(w)$	$w \cdot F_W(w)$
0	0.056314	0.056314	0.000000
1	0.187712	0.244025	0.187712
2	0.281568	0.525593	0.563135
3	0.250282	0.775875	0.750847
4	0.145998	0.921873	0.583992
5	0.058399	0.980272	0.291996
6	0.016222	0.996494	0.097332
7	0.003090	0.999584	0.021629
8	0.000386	0.999970	0.003090
9	0.000029	0.999999	0.000257
10	0.000001	1.000000	0.000010
		$\mu_W =$	**2.5**

g. $\mu_W = n \cdot p = 10 \cdot 0.25 = 2.5$ students

11. Suppose that X has a uniform distribution on the interval $[0,20]$.
  a. Graph the p.d.f. of X on the interval $[0,20]$.
  b. Graph the c.d.f. of X on the interval $[0,20]$.
  c. Find $P(X = 10)$.
  d. Find $P(X \le 10)$.
  e. Find $F_X(5)$.
  f. Find $P(0 \le X \le 20)$.
  g. Find $P(10 \le X \le 15)$ using the p.d.f.
  h. Find $P(10 \le X \le 15)$ using the c.d.f.

SOLUTION

  a.

$$\text{Height} = f_X(x)$$
$$= \frac{1}{b-a} \quad \text{for } 0 \le x \le 20$$
$$= \frac{1}{20-0}$$
$$= \frac{1}{20}$$

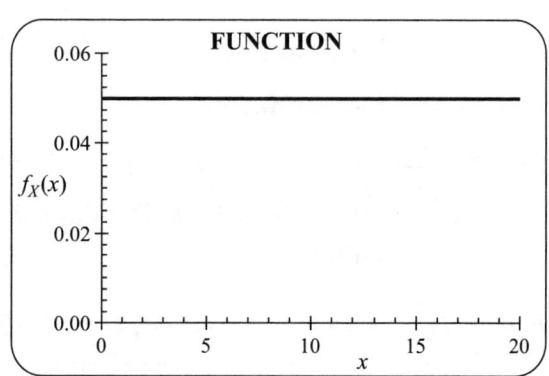

  b.

$$F_X(x) = \frac{x-a}{b-a} \quad \text{for } 0 \le x \le 20$$
$$= \frac{x-0}{20-0}$$
$$= \frac{x}{20}$$

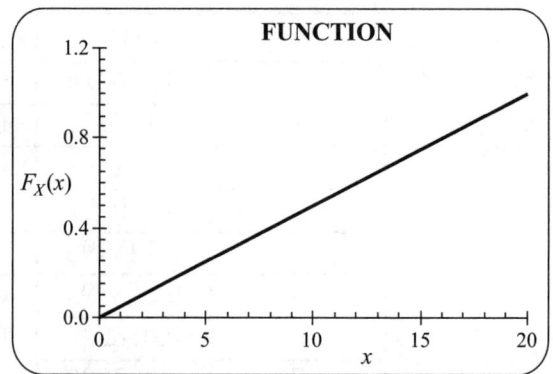

  c. $P(X = 10) = 0$

  d. $P(X \le 10) = F_X(10) = \frac{10}{20} = 0.5$

  e. $F_X(5) = \frac{5}{20} = 0.25$

f. $P(0 \leq X \leq 20)$ = area under the graph of the *p.d.f.* over the interval [0,20] = 1. This can be verified from the graph below.

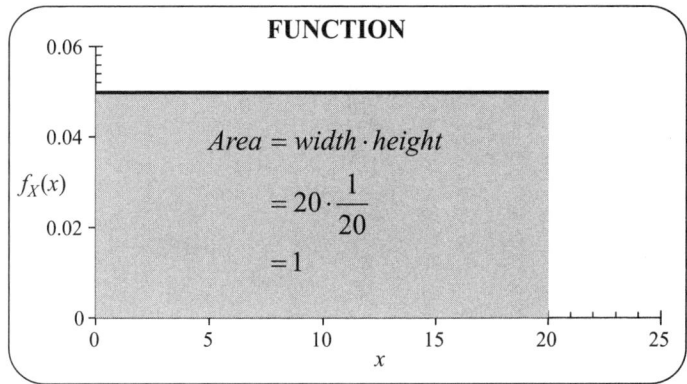

g. $P(10 \leq X \leq 15)$ = area under the graph of the *p.d.f.* over the interval [10,15]

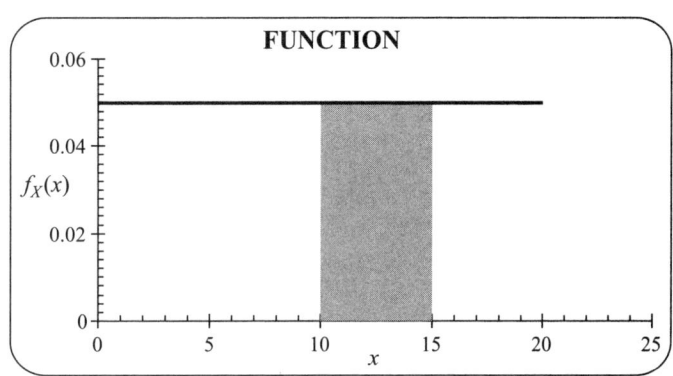

$$Area = base \cdot height$$
$$= (15-10) \cdot 0.05$$
$$= 0.25$$

h. $P(10 \leq X \leq 15) = F_X(15) - F_X(10) = 0.75 - 0.50 = 0.25$

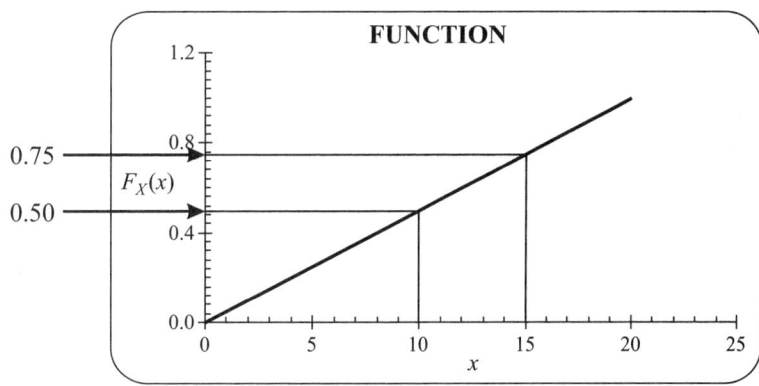

12. Let T be the length of time (in minutes) required for an assignment. Past experience shows that T has a uniform distribution on the interval [60,90].
    a. Find a formula for the *p.d.f.* of T.
    b. Use the formula for the *p.d.f.* to compute the probability that between 60 and 75 minutes will be required for the assignment.
    c. Find a formula for the *c.d.f.* of T.

d. Use the formula for the *c.d.f.* to compute the probability that between 60 and 75 minutes will be required for the assignment.

e. Find the expected length of time required for the assignment.

SOLUTION

a. $f_T(t) = \begin{cases} 0 & \text{if } t < 60 \\ \dfrac{1}{30} & \text{if } 60 \le t \le 90 \\ 0 & \text{if } t > 90 \end{cases}$

b. $P(60 \le T \le 75) = 0.5$

c. $F_T(t) = \begin{cases} 0 & \text{if } t < 60 \\ \dfrac{t-60}{30} & \text{if } 60 \le t < 90 \\ 1 & \text{if } t \ge 90 \end{cases}$

d. $P(60 \le T \le 75) = 0.5$

e. $E(T) = 75$ minutes

13. Suppose X has a uniform distribution on the interval $[4,10]$.
    a. Graph the *p.d.f.* of X on the interval $[0,15]$.
    b. Find a formula for the *p.d.f.* of X.
    c. Graph the *c.d.f.* of X on the interval $[0,15]$.
    d. Find a formula for the *c.d.f.* of X.
    e. Calculate $P(X = 6)$.
    f. Calculate $P(X \le 6)$ using the *p.d.f.*
    g. Calculate $P(X \le 6)$ using the *c.d.f.*
    h. Calculate $P(6 \le X \le 11)$ using the *p.d.f.*
    i. Calculate $P(6 \le X \le 11)$ using the *c.d.f.*

SOLUTION

a.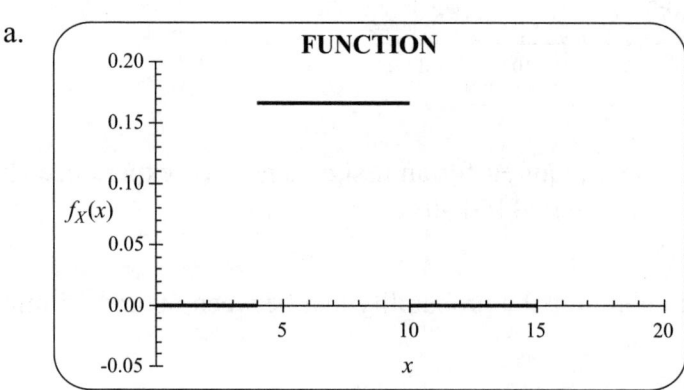

b. $f_X(x) = \begin{cases} 0 & \text{if } x < 4 \\ \dfrac{1}{6} & \text{if } 4 \leq x \leq 10 \\ 0 & \text{if } x > 10 \end{cases}$

c.

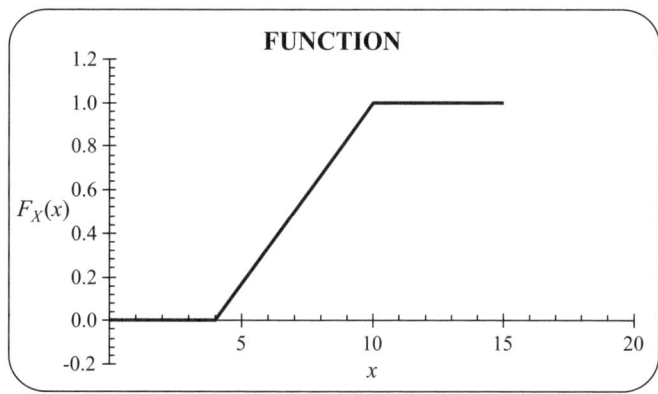

d. $F_X(x) = \begin{cases} 0 & \text{if } x < 4 \\ \dfrac{x-4}{6} & \text{if } 4 \leq x < 10 \\ 1 & \text{if } x \geq 10 \end{cases}$

e. $P(X = 6) = 0$

f. $P(X \leq 6)$ = area under the graph of f_X over the interval $[4,6]$

$P(X \leq 6) = (6-4) \cdot \dfrac{1}{6} = \dfrac{2}{6} = \dfrac{1}{3}$

g. $P(X \leq 6) = F_X(6) = \dfrac{6-4}{6} = \dfrac{1}{3}$

h. $P(6 \leq X \leq 11) = P(6 \leq X \leq 10)$ because $P(X > 10) = 0$

$P(6 \leq X \leq 10)$ = area under the graph of f_X over the interval $[6,10] = (10-6) \cdot \dfrac{1}{6} = \dfrac{4}{6} = \dfrac{2}{3}$

i. $P(6 \leq X \leq 11) = F_X(11) - F_X(6) = 1 - \dfrac{6-4}{6} = \dfrac{2}{3}$

14. Let Y be an exponential random variable with parameter $\alpha = 20$.
   a. Find a formula for the *c.d.f.* of Y.
   b. Use the *c.d.f.* to compute $P(6 \leq Y \leq 12)$.
   c. Find a formula for the *p.d.f.* of Y.
   d. Find $E(Y)$.

SOLUTION

a. $F_Y(y) = \begin{cases} 0 & \text{if } y < 0 \\ 1 - e^{-y/20} & \text{if } y \geq 0 \end{cases}$

b. $P(6 \leq Y \leq 12) = F_Y(12) - F_Y(6)$
$= (1 - e^{-12/20}) - (1 - e^{-6/20})$
$\cong 0.4512 - 0.2592$
$= 0.1920$

c. $f_Y(y) = \begin{cases} 0 & \text{if } y < 0 \\ \dfrac{1}{20} \cdot e^{-y/20} & \text{if } y \geq 0 \end{cases}$

d. $E(Y) = \alpha = 20$

15. On average, three customers per hour use the ATM in a local grocery store. Let T be the time (in minutes) between the arrivals of consecutive customers.
    a. Find a formula for the c.d.f. of T.
    b. Use the c.d.f. to find the probability that the time between the arrivals of consecutive customers is at most 15 minutes.
    c. Use the c.d.f. to find the probability that the time between the arrivals of consecutive customers is at least 10 minutes.
    d. Find $E(T)$.

SOLUTION

a. There are $\dfrac{3 \text{ customers}}{1 \text{ hour}} = \dfrac{3 \text{ customers}}{60 \text{ minutes}}$

$\dfrac{1 \text{ customer}}{\alpha \text{ minutes}} = \dfrac{3 \text{ customers}}{60 \text{ minutes}}$

$3 \cdot \alpha = 60 \text{ minutes}$

$\alpha = 20 \text{ minutes}$

$F_T(t) = \begin{cases} 0 & \text{if } t < 0 \\ 1 - e^{-t/20} & \text{if } t \geq 0 \end{cases}$

b. $P(T \leq 15) = F_T(15) = 1 - e^{-15/20} \cong 0.5276$

c. $P(T \geq 10) = 1 - F_T(10) = 1 - (1 - e^{-10/20}) \cong 0.6065$

d. $E(T) = \alpha = 20$ minutes

Probability Distributions

16. On average, 15 patients per hour seek treatment at the emergency room of a hospital in a large metropolitan area. Let T be the time (in minutes) between patient arrivals. The *p.d.f.* of T is given below.

$$f_T(t) = \begin{cases} 0 & \text{if } t < 0 \\ 0.25e^{-0.25t} & \text{if } t \geq 0 \end{cases}$$

a. Find the probability that the time between patient arrivals is exactly five minutes.
b. Find a formula for the *c.d.f.* of T.
c. Use the *c.d.f.* to find the probability that the time between patient arrivals is at least 10 minutes.
d. Find the expected value of T.

SOLUTION

a. $P(T = 5) = 0$

b. $F_T(t) = \begin{cases} 0 & \text{if } t < 0 \\ 1 - e^{-0.25t} & \text{if } t \geq 0 \end{cases}$

c. $P(T \geq 10) \cong 0.0821$

d. $E(T) = 4$ minutes

17. The *p.d.f.* and *c.d.f.* of a continuous random variable X are plotted below.

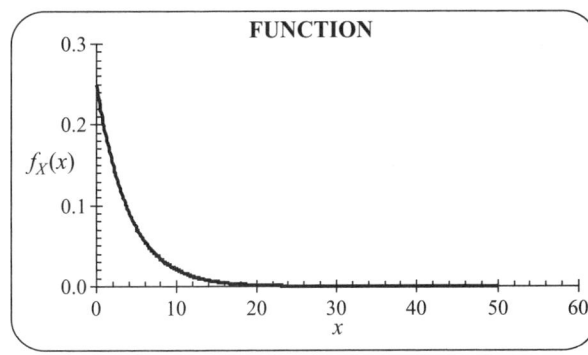
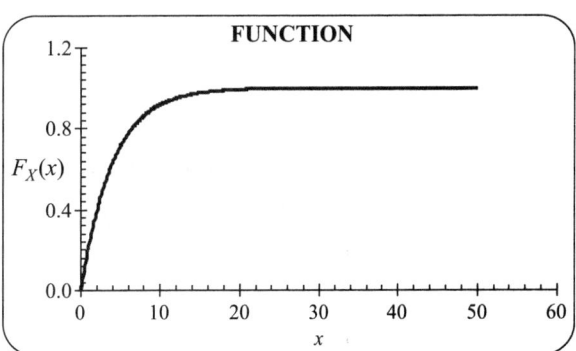

a. Shade the region of the graph of the *p.d.f.* that corresponds to $P(X \leq 10)$.
b. Use the graph of the *c.d.f.* to estimate $P(X \leq 10)$

SOLUTION

a.
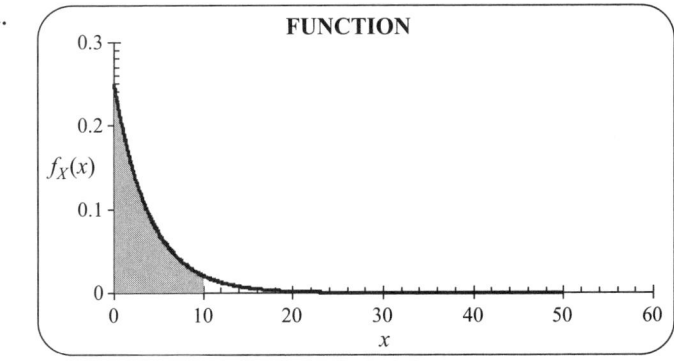

b.

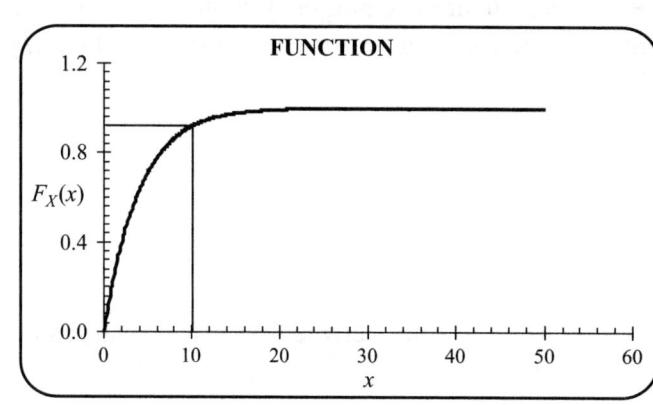

$$P(X \leq 10) = F_X(10) \cong 0.92$$

18. The *p.d.f.* and *c.d.f.* of a continuous random variable X are plotted below.

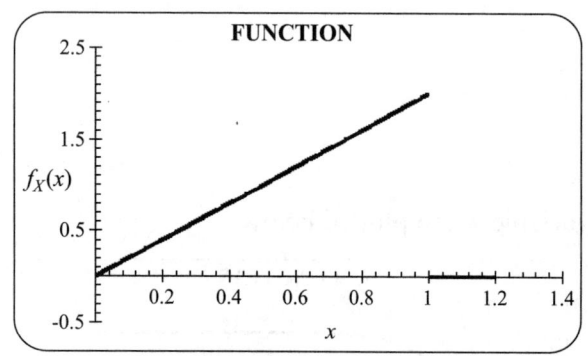 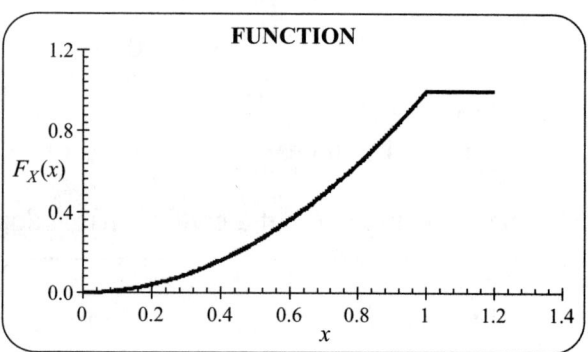

a. Shade the region of the graph of the *p.d.f.* that corresponds to $P(0.4 \leq X \leq 0.8)$.

b. Use the graph of the *c.d.f.* to estimate $P(0.4 \leq X \leq 0.8)$.

SOLUTION

a.

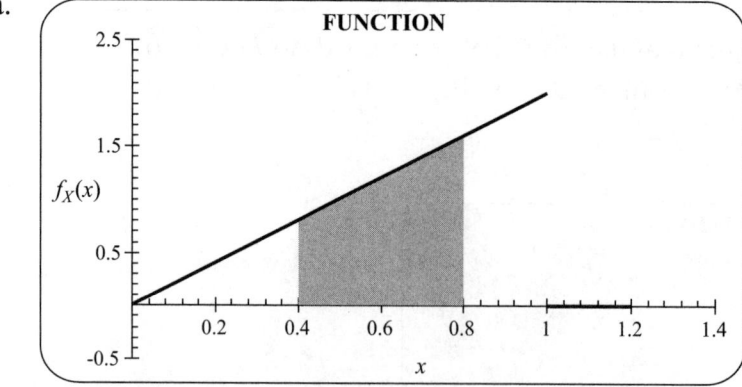

b.

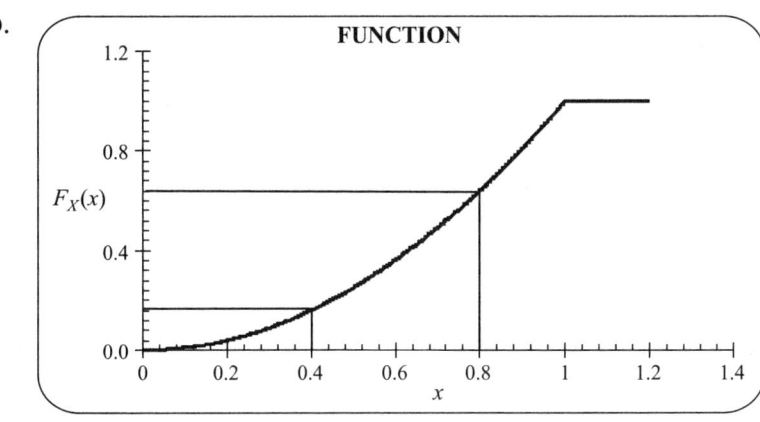

$P(0.4 \leq X \leq 0.8) = F_X(0.8) - F_X(0.4) \cong 0.64 - 0.16 = 0.48$

Random Sampling

Properties of any random variable can be estimated by looking at a sample of observations.

For example, the *p.m.f.* of a finite random variable can be approximated by a relative frequency histogram. This histogram can then be used to estimate probabilities for the random variable. Recall that the empirical interpretation of probability is that a probability is a long-term (over a large number of trials) relative frequency.

Example 1a: Jack has three pennies, four nickels, two dimes and one quarter in his pocket. Let X be the denomination of a coin selected at random from his pocket.

The *p.m.f.* of X is given by:

x (in cents)	1	5	10	25
$f_X(x)$	0.3	0.4	0.2	0.1

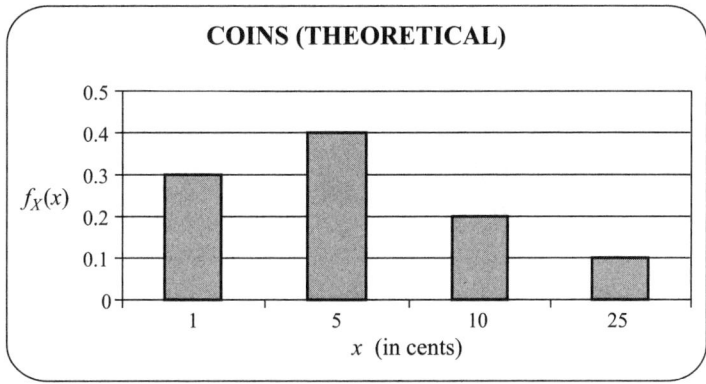

Forty observations of the random variable are given below.

10	1	5	1	5	5	1	10
1	1	5	25	10	1	5	1
5	10	1	5	1	5	10	5
5	1	10	10	5	5	1	10
25	10	1	5	10	1	25	5

Bin	Frequency	Relative Frequency
1	13	0.325
5	15	0.375
10	9	0.225
25	3	0.075

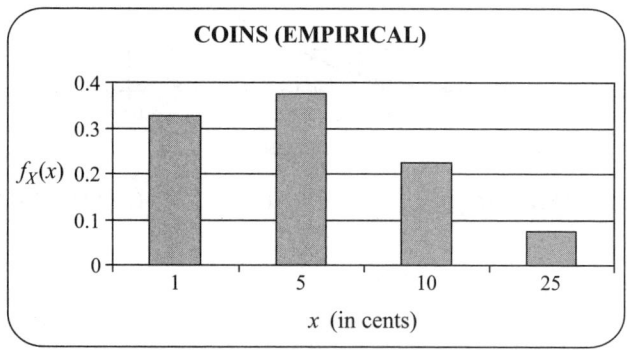

Notice that for a sample of size forty, the empirical probabilities closely approximate the theoretical probabilities. The theoretical and empirical probabilities are given in the table below.

x (in cents)	1	5	10	25
Theoretical Probability	0.30	0.40	0.20	0.10
Empirical Probability	0.325	0.375	0.225	0.075

Other information about random variables can also be obtained from the corresponding random samples. In particular, the expected value or mean of a random variable can be approximated by the sample mean.

Example1b: The expected value or mean of X, the denomination of a coin selected at random from Jack's pocket, is given by

$$E(X) = \mu_X = 1 \cdot 0.3 + 5 \cdot 0.4 + 10 \cdot 0.2 + 25 \cdot 0.1 = 6.8 \text{ cents.}$$

In other words, we would expect to obtain 6.8 cents on average in a large number of repetitions of this experiment. The sample mean for the forty observations given above is

$$\bar{x} = \frac{1}{n} \sum_{i=1}^{n} x_i$$
$$= \frac{1}{40}(10 + 1 + 5 + \cdots + 1 + 25 + 5)$$
$$= 6.235 \text{ cents.}$$

This value differs from the theoretical value by only 0.565 cents.

Example 2: A-Tech purchased a robotic assembly line to help streamline its manufacturing of circuit boards. According to the specifications of the robotic assembly line, eight out of every one hundred circuit boards produced are defective. Let $D = 0$ if the circuit board is defective and let $D = 1$ if the circuit board is not defective.

a. Find the *p.m.f.* of *D*.
b. Graph the *p.m.f.*
c. Compute *E(D)* and interpret the results.

Thirty observations of the random variable are given below.

1	1	1	1	1	1
1	1	1	1	1	1
1	1	1	1	1	1
1	1	1	1	1	1
0	1	1	0	1	1

d. Compute the sample mean.
e. Sort the data into bins and graph the estimated *p.m.f.*
f. How close is the empirical probability that a circuit board will not be defective to the theoretical probability?

SOLUTION

a. Recall this is a Bernoulli random variable (refer to page 135).

d	$f_D(d)$
0	0.08
1	0.92

b.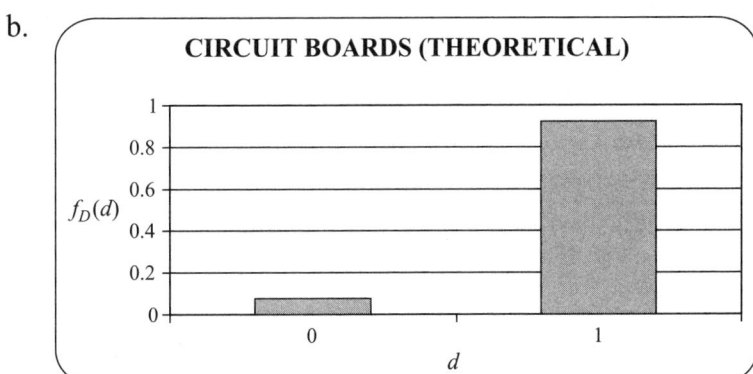

c. $E(D) = 0 \cdot 0.08 + 1 \cdot 0.92 = 0.92$. On average, for a large number of circuit boards, the probability that a circuit board will not be defective is 92%.

d. $\bar{d} = \frac{1}{n}\sum_{i=1}^{n} d_i$

$= \frac{1}{30}(1+1+1+\cdots+0+1+1)$

$\cong 0.933$

e.

Bin	Frequency	Relative Frequency
0	2	0.067
1	28	0.933

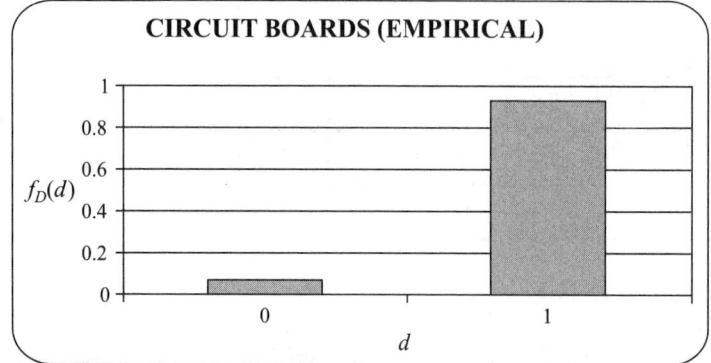

f. Difference = $\bar{d} - E(D)$ = 0.933 − 0.920 = 0.013 = 1.13%

Example 3: Marcela, the owner of the Veggie Store, has chosen a new vendor for oranges. She just received her first shipment of individual bags and wants to know how accurate the weight stamped on each bag is; the weight of each bag is marked as three pounds. Marcela randomly selects eight bags and weighs them. The results are given below.

Bag	1	2	3	4	5	6	7	8
Weight (in pounds)	3.2	2.7	3.1	2.9	3.2	2.7	3.3	2.8

Let R be the random variable that represents the observed weight of a bag of oranges and assume that her machine is calibrated correctly.

a. Marcela is very customer-oriented and wants to know the probability that a customer will receive a bag of oranges that is less than the marked weight of three pounds.

b. Does it appear that the bags of oranges do in fact weigh three pounds on average?

SOLUTION

a. There are four bags (2, 4, 6, & 8) that weigh less than three pounds. Therefore, $P(R \leq 3) \cong$ 4/8 = 0.5, and Marcela assumes that approximately half of all bags of oranges from the new vendor weigh less than three pounds. For this reason, Marcela contacted her vendor in order to improve the situation. It turns out that the vendor's scale had not been calibrated in two years. After it was recalibrated, only 5% of the bags weighed less than three pounds.

b. $\bar{r} = \dfrac{1}{8}(3.2 + 2.7 + 3.1 + 2.9 + 3.2 + 2.7 + 3.3 + 2.8)$

= 2.9875 pounds

The sample mean and expected value differ by 0.0125 pounds.

Random Sampling

Example 4: The washers and dryers at the self-serve laundromat, Super Suds, operate on coins: quarters, dimes, and nickels. Occasionally the coins jam inside a machine and the customer loses money. Let D be the denomination of money that a customer loses using the dryers in Super Suds during business hours and let W be the denomination of money that a customer loses using the washers during business hours. The p.m.f.'s of D and W are given below.

d (in dollars)	0.05	0.10	0.25
$f_D(d)$	0.10	0.60	0.30

w (in dollars)	0.05	0.10	0.25
$f_W(w)$	0.15	0.25	0.60

a. In which type of machine can a customer expect to lose more money?

b. Ten observations of D and ten observations of W were obtained in order to tabulate the amount of money in each machine. Since the machines automatically keep track of how many times they are used, it is easy for the attendant to know how much extra money is in each machine. These observations are as follows:

Observations of D: 0.10, 0.25, 0.05, 0.10, 0.10, 0.25, 0.10, 0.05, 0.25, & 0.25
Observations of W: 0.05, 0.25, 0.25, 0.25, 0.10, 0.10, 0.05, 0.25, 0.05, & 0.25

Are customers losing more or less money than expected?

SOLUTION

a. $\mu_D = 0.05 \cdot 0.10 + 0.10 \cdot 0.60 + 0.25 \cdot 0.30 = 0.14$
$\mu_W = 0.05 \cdot 0.15 + 0.10 \cdot 0.25 + 0.25 \cdot 0.60 = 0.1825$
A customer can expect to lose more money using the washing machines.

b. $\bar{d} = \dfrac{1}{10}(0.10+0.25+0.05+0.10+0.10+0.25+0.10+0.05+0.25+0.25) = 0.15$

$\bar{w} = \dfrac{1}{10}(0.05+0.25+0.25+0.25+0.10+0.10+0.05+0.25+0.05+0.25) = 0.16$

For the dryers, the customers are losing more money than expected because the difference between $\bar{d}$ and μ_D is 0.01. The customers are losing less money than expected using the washers because the difference between $\bar{w}$ and μ_W is -0.02.

Example 5: Ten students at Pima College were surveyed and asked how many hours they study during the week. The results were: 16, 15, 25, 10, 18, 5, 8, 14, 20, 14. Let X represent the number of hours a student spends studying during the week. Estimate the expected value, $E(X)$, and the probability that a student spends fewer than ten hours a week studying.

SOLUTION

$\bar{x} \cong E(X) = \dfrac{1}{10}(16+15+25+10+18+5+8+14+20+14) = 14.5$ hours

$P(X<10) \cong 2/10 = 0.20$

At the beginning of this section we discussed that the p.m.f. of a finite random variable can be approximated by a relative frequency histogram, which then can be used to estimate probabilities for the random variable.

We can also approximate the *p.d.f.* of a continuous random variable with an adjusted relative frequency histogram. This histogram can then be used to estimate probabilities for the random variable. Recall that that $P(a \leq X \leq b)$ is equal to the area between the graph of the *p.d.f.* and the horizontal axis over the interval $[a, b]$. Therefore, if we want to use a histogram to approximate the *p.d.f.*, then the area of each rectangle must be equal to the relative frequency (probability) of the corresponding bin. The area of a rectangle is the product of the width and the height. Since we know the area, which represents the probability, and we also know the bin width, we can solve for the height.

$$area = width \cdot height$$

$$height = \frac{area}{width}$$

The area of each rectangle represents the probability. Note that if the bin width is 1, then the area of each rectangle is equal to the height.

For example, the bar below is from a histogram that has a bin width of 3. The relative frequency of this bin is 0.12 (or 12%). We want 0.12 to represent the area of the rectangle. Therefore, we must divide the relative frequency by the bin width to obtain the height. Sometimes the height is referred to as the **adjusted relative frequency**.

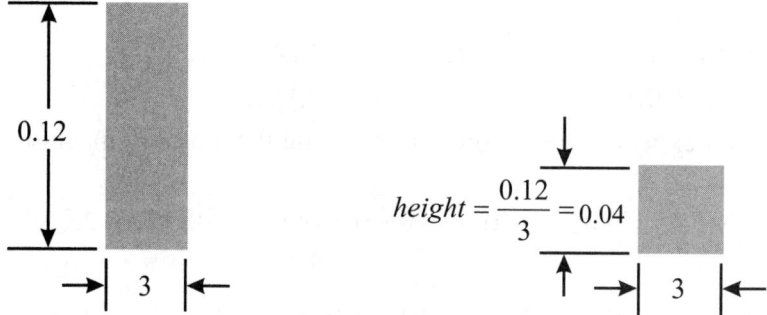

The area of the rectangle on the right is $0.04 \cdot 3 = 0.12$, and the height of the rectangle on the left is 0.12. This height is the relative frequency of the corresponding bin.

Finally, a histogram can be used to determine the type of distribution of the underlying random variable. Some examples are given below.

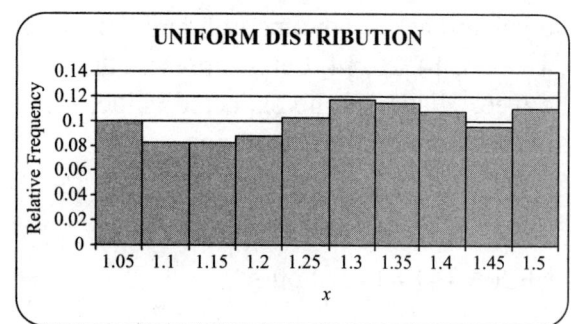
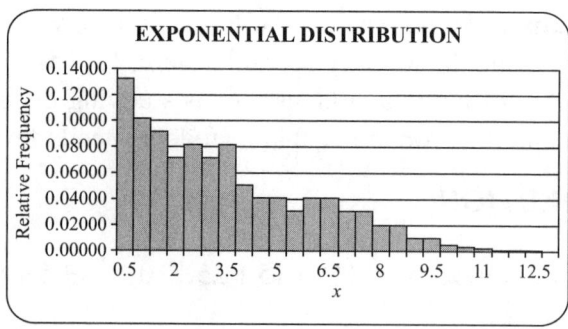

Random Sampling

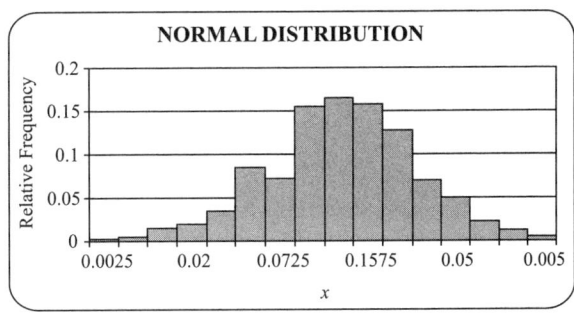

Example 6: One hundred students were selected at random from those taking an exam. The length of time (in minutes) that it took each student to complete the exam was recorded. The results of the random sample are given in the table below.

43.82	42.45	43.72	49.52	40.74	40.05	47.56	45.29	49.86	46.72
41.01	40.45	43.56	40.53	41.98	49.26	46.27	47.97	49.26	47.32
45.96	40.32	49.10	47.05	40.64	41.00	41.74	48.06	49.04	45.85
48.99	41.64	44.66	48.17	43.58	42.57	44.05	42.62	45.45	41.52
48.85	42.20	44.26	49.73	44.87	47.76	45.52	41.78	45.01	48.92
49.58	40.17	43.04	44.66	45.11	46.80	47.12	48.67	46.75	43.78
40.14	42.85	49.76	43.00	43.73	48.09	45.55	41.15	44.90	42.00
44.07	43.43	48.07	47.50	49.86	47.24	41.81	40.60	41.46	42.06
48.63	45.54	49.91	43.51	40.41	40.85	49.70	47.62	40.38	43.34
41.39	43.57	42.56	47.76	42.31	41.32	46.87	47.38	47.96	43.25

The histogram shown below was constructed from the sample data.

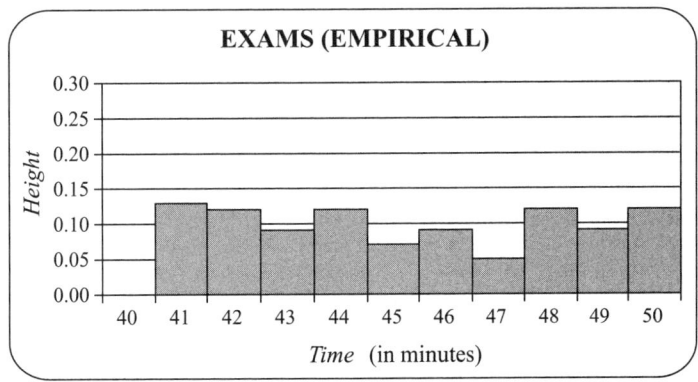

From this histogram, it appears that T, the time (in minutes) required for the exam, has a uniform distribution on the interval [40,50]. The graph of the assumed *p.d.f.* of T is given below.

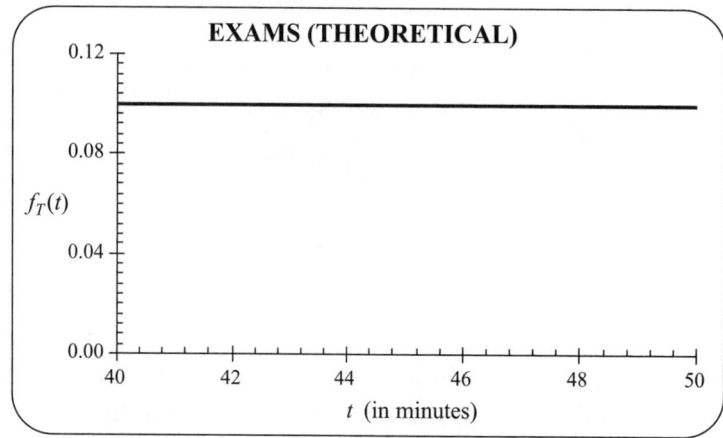

Notice that the general shape of the empirical distribution is similar to that of the theoretical distribution. A larger sample would enhance the similarities between the distributions. We can now use the assumed *p.d.f.* to approximate probabilities for T. For example, the probability that it will take a randomly selected student between 42.6 and 45.7 minutes to complete the exam is approximately given by $0.10 \cdot (45.7 - 42.6)$ or 0.31.

As is the case for finite random variables, the expected value of a continuous random variable can be approximated by the sample mean. The sample mean for this example is as follows:

$$\bar{t} = \frac{1}{n}\sum_{i=1}^{n} t_i = \frac{1}{100}(43.82 + 41.01 + 45.96 + \cdots + 42.06 + 43.34 + 43.25) = 44.85 \text{ minutes.}$$

If T has a uniform distribution on the interval [40, 50], then the expected value of T is given by

$$E(T) = \mu_T = \frac{a+b}{2} = \frac{40+50}{2} = 45 \text{ minutes.}$$

The difference between the sample and theoretical means is only 0.15.

Once again it is very important to understand that the difference between the theoretical and sample means will decrease as the sample size becomes larger. Remember the saying "the more the merrier"; in statistics "the more data the merrier."

Example 7: The estimated *p.m.f.* of a finite random variable Y is given below.

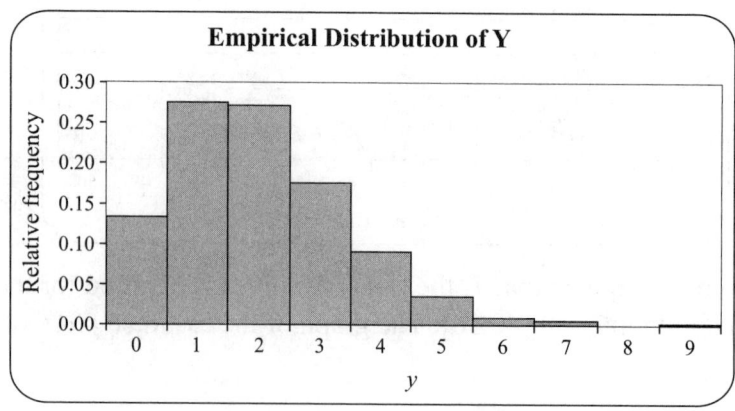

Although the theoretical distribution of Y is unknown, we can approximate probabilities and the expected value for Y from the sample. For example, the probability that Y is equal to 4 can be approximated by the relative frequency of 4 in the sample of 500 observations. In other words, $P(Y = 4) \cong 0.088$. The expected value or mean of Y can be approximated by the mean of the sample of 500 observations, which is given by $E(Y) \cong \bar{y} = 1.996$. Since the theoretical probabilities and expected value are unknown, the probabilities and expected value approximated by the sample are only estimates of the true values. The accuracy of the estimates improves as the sample size increases.

Example 8: One hundred observations of a continuous random variable X are shown in the table below.

1.43	9.01	2.22	3.35	3.48	6.92	4.62	3.30	0.47	1.31
0.35	7.45	4.14	13.23	3.49	4.52	0.60	0.86	1.79	5.74
7.38	0.17	1.87	6.42	5.28	1.96	1.80	0.20	0.97	5.96
0.75	2.40	1.51	0.22	7.22	8.73	8.81	0.48	0.64	1.95
0.30	2.04	2.85	4.38	1.85	2.14	0.93	3.12	0.95	0.03
3.66	3.05	3.14	5.39	4.28	6.71	5.17	4.32	4.59	13.07
0.16	1.73	0.15	2.75	8.51	1.46	12.84	0.72	1.24	5.05
1.14	2.54	2.52	1.92	0.51	4.44	6.09	1.35	17.85	2.29
0.47	5.99	3.76	0.47	3.10	2.13	6.90	5.57	1.35	5.24
2.11	0.34	5.05	2.48	5.99	4.57	6.43	6.01	0.81	3.45

a. Construct a histogram that would approximate the *p.d.f.* of the random variable X.
b. This histogram appears to approximate what type of continuous random variable?
c. Use the sample data to compute an estimate of the expected value, $E(X)$.
d. What is the equation of the assumed *p.d.f.*?
e. What is the equation of the assumed *c.d.f.*?
f. Use the formula for the assumed *c.d.f.* to compute an estimate of $P(5 \leq X \leq 10)$.
g. Use the formula for the assumed *c.d.f.* to compute an estimate of $P(X \geq 9)$.

SOLUTION

a. The adjusted relative frequency is equal to the relative frequency divided by the bin width.

Bin	Frequency	Relative Frequency	Adjusted Relative Frequency
2	39	0.39	0.195
4	23	0.23	0.115
6	20	0.20	0.100
8	10	0.10	0.050
10	4	0.04	0.020
12	0	0.00	0.000
14	3	0.03	0.015
16	0	0.00	0.000
18	1	0.01	0.005

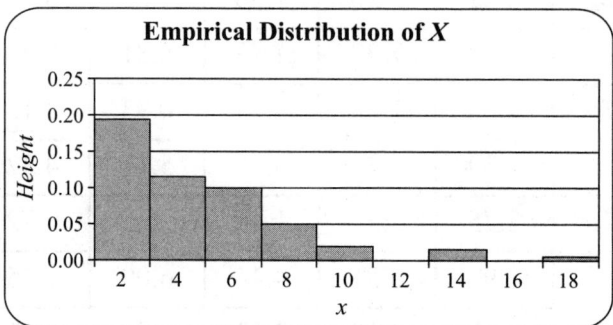

b. The shape of the histogram in Part a is similar to the shape of the *p.d.f.* shown below. Furthermore, this is the graph of an exponential random variable which was discussed in the section ***Probability Distributions***. Therefore, it would be reasonable to assume that X is an exponential random variable.

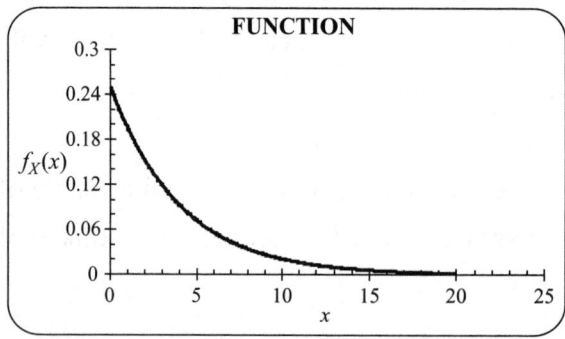

c. $E(X) \cong \bar{x} = \dfrac{1}{100}(1.43 + 0.35 + 7.38 + \cdots + 2.29 + 5.24 + 3.45) = 3.62$

d. $f_X(x) \cong \begin{cases} 0 & \text{if } x < 0 \\ \dfrac{1}{3.62} e^{-x/3.62} & \text{if } x \geq 0 \end{cases}$

e. $F_X(x) \cong \begin{cases} 0 & \text{if } x < 0 \\ 1 - e^{-x/3.62} & \text{if } x \geq 0 \end{cases}$

f. $P(5 \leq X \leq 10) = F_X(10) - F_X(5)$
$= \left(1 - e^{-10/3.62}\right) - \left(1 - e^{-5/3.62}\right)$
$\cong 0.9369 - 0.7487$
$= 0.1882$

g. $P(X \geq 9) = 1 - P(X \leq 9)$
$= 1 - \left(1 - e^{-9/3.62}\right)$
$\cong 1 - 0.9168$
$= 0.0832$

Exercises

1. The p.m.f. of a finite random variable R is given below.

r	0	5	10	15
$f_R(r)$	0.4	0.3	0.2	0.1

   Forty observations of R are given in the table below.

10	0	15	0	5	5	10	0
5	5	0	0	10	0	0	0
5	5	10	5	10	5	0	10
5	0	5	0	5	0	10	5
5	10	5	15	10	0	0	15

   a. Find μ_R.
   b. Find $\bar{r}$.

   SOLUTION

   a. $\mu_R = 5$

   b. $\bar{r} = 5.125$

2. Five observations of V, the volume (in millions of shares) of the New York Stock Exchange, are: 731, 542, 610, 689, & 558.
   a. Use this information to estimate $P(V \geq 675)$.
   b. Use this information to estimate $E(V)$.

SOLUTION

a. $P(V \geq 675) \cong 2/5 = 0.4$

b. $E(V) \cong \bar{v} = 626$ million shares

3. Twelve students were asked how much money they spend on textbooks per semester. The results were: $200, $175, $450, $300, $350, $250, $150, $200, $320, $370, $400, & $250.

    a. Use this information to estimate the probability that a randomly selected student spends less than $275 in a given semester.
    b. Use this information to estimate the probability that a randomly selected student spends more than $325 in a given semester.
    c. Use this information to estimate the average amount of money spent on textbooks per semester.

SOLUTION

a. 0.50

b. 1/3

c. $284.58

4. Forty observations of a continuous random variable X are shown in the table below.

7.07	4.87	4.27	3.18	7.60	4.40	6.17	2.27
7.02	6.69	7.71	2.15	6.70	7.25	2.58	6.03
5.03	7.86	3.40	3.23	3.64	7.85	2.61	6.10
5.46	7.52	5.00	4.23	2.76	5.88	5.18	5.81
3.92	7.96	2.48	2.64	7.07	5.77	4.74	4.66

a. Construct a histogram that would approximate the *p.d.f.* for the random variable X.
b. This histogram would approximate what type of continuous random variable?
c. What is the equation of the assumed *p.d.f.*?
d. What is the equation of the assumed *c.d.f.*?
e. Use the formula for the assumed *c.d.f.* to compute an estimate of $P(3 \leq X \leq 7)$.
f. Use the formula for the assumed *c.d.f.* to compute an estimate of $P(X \geq 3)$.
g. Use the sample data to compute an estimate of the expected value, $E(X)$.

SOLUTION

a.

Bin	Frequency	Relative Frequency	Adjusted Relative Frequency
2.0	0	0.00	0.00
2.6	4	0.10	0.17
3.2	4	0.10	0.17
3.8	3	0.08	0.13
4.4	4	0.10	0.17
5.0	4	0.10	0.17
5.6	3	0.08	0.13
6.2	6	0.15	0.25
6.8	2	0.05	0.08
7.4	4	0.10	0.17
8.0	6	0.15	0.25

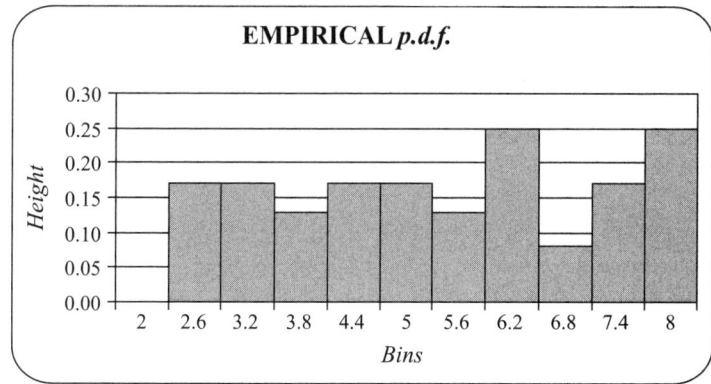

b. From this histogram, it appears that X, has a uniform distribution on the interval $[2, 8]$. The graph of the assumed $p.d.f.$ of X is given below.

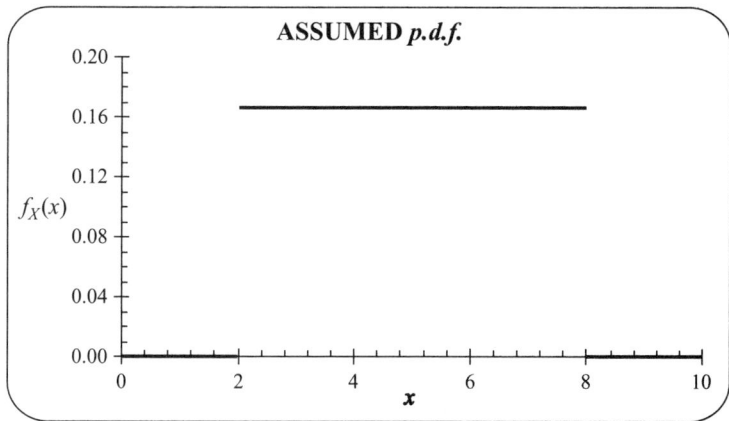

c. $f_X(x) = \begin{cases} 0 & \text{if } x < 2 \\ \dfrac{1}{6} & \text{if } 2 \leq x \leq 8 \\ 0 & \text{if } x > 8 \end{cases}$

d. $F_X(x) = \begin{cases} 0 & \text{if } x < 2 \\ \dfrac{x-2}{6} & \text{if } 2 \leq x < 8 \\ 1 & \text{if } x \geq 8 \end{cases}$

e. $P(3 \leq X \leq 7) \cong 0.667$

f. $P(X \geq 3) \cong 0.883$

g. $E(X) \cong \bar{x} \cong 5.17$

5. The *p.m.f.* of a finite random variable X is given below.

x	2	4	6	8	10
$f_X(x)$	0.2	0.1	0.3	0.35	0.05

Nine observations of X are: 6, 2, 6, 8, 8, 10, 2, 6, & 8.

a. Find μ_X.

b. Find $\bar{x}$.

SOLUTION

a. $\mu_X = 5.9$

b. $\bar{x} = 6.22$

6. Fill in the missing values for the Adjusted Relative Frequency column.

Bin	Frequency	Relative Frequency	Adjusted Relative Frequency
60	1	0.0200	
68	10	0.2000	
76	15	0.3000	
84	16	0.3200	
92	7	0.1400	
100	1	0.0200	

Random Sampling

SOLUTION

Bin	Frequency	Relative Frequency	Adjusted Relative Frequency
60	1	0.0200	0.0025
68	10	0.2000	0.0250
76	15	0.3000	0.0375
84	16	0.3200	0.0400
92	7	0.1400	0.0175
100	1	0.0200	0.0025

7. The table above summarizes 50 observations of a continuous random variable W.
   a. Sketch a histogram that represents the estimated *p.d.f.*
   b. Based upon the histogram, what type of random variable do you believe W is? (Hint: refer to the histograms on pp. 172–173.)
   c. What is the difference between the frequency and the relative frequency?
   d. What is the difference between the relative frequency and the adjusted relative frequency?

SOLUTION

   a. Left for the student to graph.
   b. Normal random variable.
   c. The frequency is the number of observations that fall within a particular bin range. For example, in the table above, there are 10 values of data that are in the bin (60, 68]. The relative frequency is the frequency divided by the number of observations in the sample. So the relative frequency of the bin (60, 68] is 10/50 = 0.20.
   d. Refer to Part c for the explanation of relative frequency. If a histogram is used to estimate a *p.d.f.*, the area of each rectangle must be equal to the relative frequency of the corresponding bin. This is accomplished by dividing the relative frequency by the bin width. This is shown in the illustrations below for the bin (60, 68].

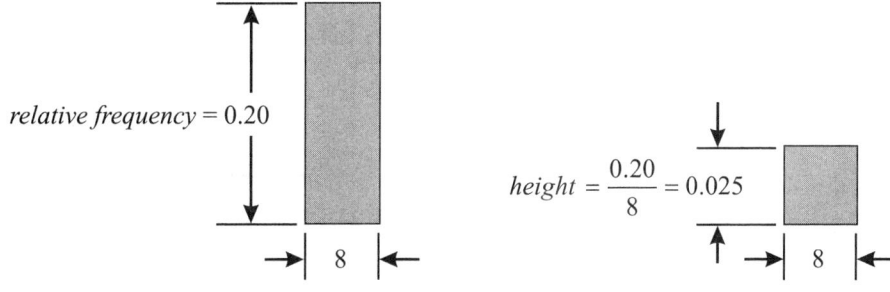

Simulation

Two functions in *Excel* can be used to generate random numbers: **=RANDBETWEEN(*a*,*b*)** is used to generate random integers between *a* and *b*, and **=RAND()** is used to generate random numbers between 0 and 1. These functions may also be combined with other functions in order to do more complicated simulations.

Example 1: Use the **RANDBETWEEN** function in *Excel* to simulate 60 rolls of a fair tetrahedral die.

SOLUTION

A tetrahedral die has four faces, numbered 1, 2, 3, and 4. We select the **RANDBETWEEN** function from the *Math & Trig* category and enter the information shown below.

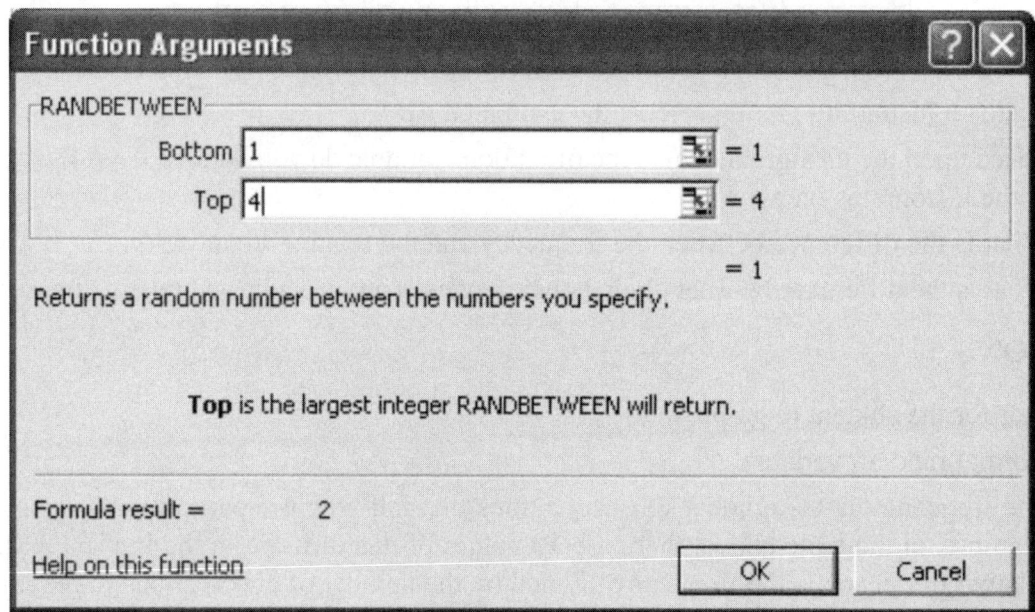

This formula returns the outcome of the first roll. Then we drag the formula five columns to the right and nine rows down to obtain the outcomes of the remaining 59 rolls. The results are given to the right.

Roll					
2	1	4	1	3	2
1	3	1	3	1	1
3	4	3	4	4	2
4	1	4	2	1	3
3	2	3	1	3	2
2	1	1	2	1	3
3	4	2	4	3	4
1	2	4	2	1	2
2	4	3	4	2	3
4	2	1	4	3	1

Simulation

Example 2: Use the **RAND** and **IF** functions in *Excel* to simulate 60 rolls of a fair tetrahedral die.

SOLUTION

We use the **RAND** function to generate 60 random numbers between 0 and 1 and then use a nested **IF** statement to record the outcomes of the rolls. The **RAND** function is found in the *Math & Trig* category.

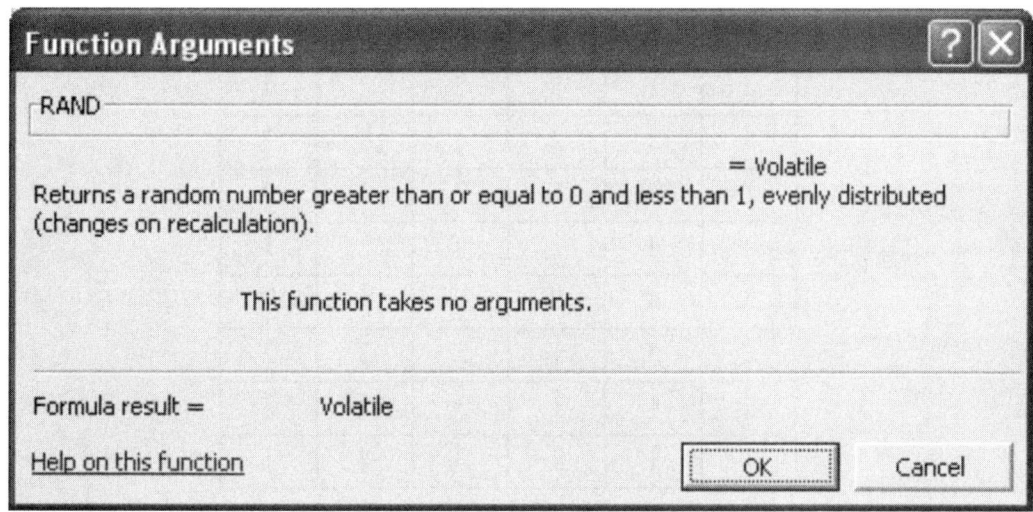

This formula returns one random number. Then we drag the formula five columns to the right and nine rows down to obtain 59 additional random numbers. The results are given below.

Random Numbers					
0.880560	0.886536	0.663647	0.148402	0.934013	0.018402
0.376215	0.213233	0.830466	0.934489	0.691246	0.352559
0.284405	0.197645	0.727725	0.471695	0.290323	0.789643
0.912390	0.680756	0.364267	0.725807	0.674118	0.294111
0.034119	0.518666	0.141972	0.380504	0.595578	0.443578
0.861545	0.901706	0.153619	0.761729	0.277523	0.204059
0.684145	0.823895	0.304105	0.310059	0.672345	0.383715
0.275855	0.737813	0.089250	0.482423	0.393572	0.433783
0.471971	0.087289	0.225117	0.698301	0.179461	0.507481
0.124050	0.002740	0.612136	0.654862	0.756937	0.049887

Next we use a nested **IF** statement to record the outcome of the roll for each random number. The four possible outcomes are equally likely, so we divide the interval from 0 to 1 into four equal parts and assign one value to each of the four subintervals. The value "1" is assigned to the interval from 0 to 0.25, the value "2" is assigned to the interval from 0.25 to 0.50, the value "3" is assigned to the interval from 0.50 to 0.75, and the value "4" is assigned to the interval from 0.75 to 1. For example, the outcome of the first roll is "4" because 0.880560, the first random number generated, is within the interval to which the value "4" is assigned. The *Excel* formula and results are given below.

=IF(*cell reference*<0.25,1,IF(*cell reference*<0.50,2,IF(*cell reference*<0.75,3,4)))

Rolls					
4	4	3	1	4	1
2	1	4	4	3	2
2	1	3	2	2	4
4	3	2	3	3	2
1	3	1	2	3	2
4	4	1	4	2	1
3	4	2	2	3	2
2	3	1	2	2	2
2	1	1	3	1	3
1	1	3	3	4	1

Example 3: The time X (in minutes) between departures of shuttle busses from the Disneyland parking lot is an exponential random variable with parameter $\alpha = 4$ minutes. Use the inverse *c.d.f.* of X and the **RAND** function in *Excel* to simulate 15 observations of X.

SOLUTION

The *c.d.f.* is a function from the real numbers to the interval [0, 1]. If we restrict the domain to the interval [0, ∞), then the restricted function is one-to-one. Thus, the inverse exists and is a function from the interval [0, 1] to the interval [0, ∞). In other words, the inverse is a function from the values of the *c.d.f.* to the possible values of X. We can use the **RAND** function in *Excel* to randomly generate a value of the *c.d.f.* Then we can use the inverse *c.d.f.* to find the corresponding value of X.

The *c.d.f.* of X is given by

$$F_X(x) = \begin{cases} 0 & \text{if } x < 0 \\ 1 - e^{-x/4} & \text{if } x \geq 0. \end{cases}$$

We set $y = F_X(x) = 1 - e^{-x/4}$ for $x \geq 0$ and solve for x. The steps are shown below.

$$y = 1 - e^{-x/4} \quad \text{Subtract 1 from both sides.}$$
$$y - 1 = -e^{-x/4} \quad \text{Multiply both sides by } -1.$$
$$1 - y = e^{-x/4} \quad \text{Take the natural logarithm of both sides.}$$
$$\ln(1 - y) = -x/4 \quad \text{Multiply both sides by 4.}$$

Simulation

$$4\ln(1-y) = -x \quad \text{Multiply both sides by } -1.$$
$$-4\ln(1-y) = x$$

Graphically, this is equivalent to computing the value on the horizontal axis of the graph of $F_X(x)$ that corresponds to a randomly selected value on the vertical axis.

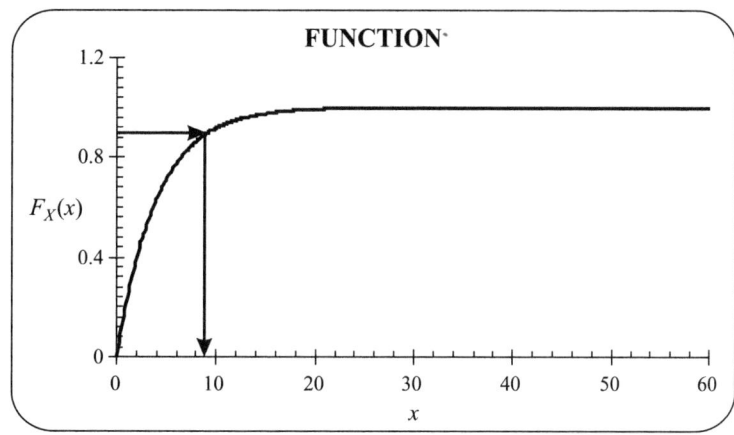

We enter the formula =-4*ln(1-RAND()) into one cell of our worksheet. Then we drag the formula four columns to the right and two columns down to obtain the remaining 14 observations. The results are given below.

Times Between Disneyland Shuttle Busses				
8.94	7.04	3.62	1.08	5.42
6.61	4.56	3.75	0.51	2.26
9.79	1.60	3.54	2.44	0.68

Example 4: The time T (in minutes) between the arrivals/departures of planes at Phoenix Sky Harbor Airport is an exponential random variable with parameter $\alpha = 1.2$ minutes. Use the inverse c.d.f. of T and the **RAND** function in *Excel* to simulate 50 observations of T.

SOLUTION

The c.d.f. of T is given by

$$F_T(t) = \begin{cases} 0 & \text{if } t < 0 \\ 1 - e^{-t/1.2} & \text{if } t \geq 0. \end{cases}$$

We set $y = F_T(t) = 1 - e^{-t/1.2}$ for $t \geq 0$ and solve for t.

$$y = 1 - e^{-t/1.2}$$
$$y - 1 = -e^{-t/1.2}$$
$$1 - y = e^{-t/1.2}$$

$$\ln(1-y) = -t/1.2$$
$$1.2 \cdot \ln(1-y) = -t$$
$$-1.2 \cdot \ln(1-y) = t$$

We enter the formula =-1.2*ln(1-RAND()) into one cell of our worksheet. Then we drag the formula four columns to the right and nine columns down to obtain the remaining 49 observations. The results are given below.

Times Between Arrivals/Departures of Planes				
0.64	0.05	1.81	1.70	0.64
1.20	0.56	0.46	2.24	0.45
2.23	0.00	0.12	0.03	0.12
0.97	0.17	0.17	1.80	0.39
0.13	0.73	1.45	2.79	3.18
0.41	2.51	0.37	1.15	1.89
2.06	0.37	2.06	3.14	1.76
0.36	2.45	0.82	1.21	0.45
1.19	0.56	1.42	0.52	1.20
1.71	0.96	0.93	0.94	0.70
2.66	0.55	0.90	1.95	2.57

Exercises

1. Let X be an exponential random variable with parameter $\alpha = 10$.
   a. Find the inverse c.d.f. of X.
   b. Use the inverse c.d.f. of X and the **RAND** function in *Excel* to simulate six observations of X.

SOLUTION

a. $x = -10 \cdot \ln(1-y)$ where $y = 1 - e^{-x/10}$
b. Results will vary. An example is given below.

2.88	5.80	8.84	2.56	16.95	8.25

Formula: =-10*ln(1-RAND())

2. The time T (in minutes) between the arrivals of customers at the ATM in a local grocery store is an exponential random variable with parameter $\alpha = 20$ minutes.

a. Find the inverse c.d.f. of T.
b. Use the inverse c.d.f. of T and the **RAND** function in *Excel* to simulate 24 observations of T.

SOLUTION

a. $t = -20 \cdot \ln(1 - y)$ where $y = 1 - e^{-t/20}$
b. Results will vary. An example is given below.

68.43	1.91	42.03	7.18	2.84	29.22
8.66	12.72	4.20	14.20	20.99	18.62
14.27	19.52	8.15	20.78	20.44	25.99
4.59	4.86	25.19	3.11	1.95	28.60

Formula: =-20*ln(1-RAND())

3. The time T (in minutes) between the arrivals of patients seeking treatment in the emergency room of a hospital in a large metropolitan area is an exponential random variable with parameter $\alpha = 4$.
   a. Find the inverse c.d.f. of T.
   b. Use the inverse c.d.f. of T and the **RAND** function in *Excel* to simulate 15 observations of T.

SOLUTION

a. $t = -4 \cdot \ln(1 - y)$ where $y = 1 - e^{-t/4}$
b. Results will vary. An example is given below.

19.91	0.39	2.77	1.04	5.20
5.30	0.33	0.15	20.22	4.06
1.84	6.35	3.92	1.75	6.80
6.43	6.53	8.33	1.44	19.24

Formula: =-4*ln(1-RAND())